Recent Trends in INVERTEBRATES

VOL. 6

H. S. Bhamrah
Kavita Juneja

ANMOL PUBLICATIONS

ANMOL PUBLICATIONS
4378/4B, Ansari Road,
Darya Ganj, New Delhi-110002.

Recent Trends in Invertebrates

ISBN 81-7041-473-3 (Set)
ISBN 81-7041-474-1 (Vol. I)
ISBN 81-7041-475-X (Vol. II)
ISBN 81-7041-476-8 (Vol. III)
ISBN 81-7041-477-6 (Vol. IV)
ISBN 81-7041-478-4 (Vol. V)
ISBN 81-7041-479-2 (Vol. VI)
ISBN 81-7041-480-6 (Vol. VII)
ISBN 81-7041-481-4 (Vol. VIII)

PRINTED IN INDIA

Published by J. L. Kumar for Anmol Publications, New Delhi and Printed at Sangita Printers, Maujpur, Delhi-53.

PREFACE

The present title has been designed to meet the requirements of the undergraduate students. In the Recent Trends in Invertebrate all phyla have been discussed. The representative types of each phylum have been described in detail. This is followed by the classification of the phylum with a description of its common animals. Recent findings and work have been incorporated in the text. Special care has been taken to make the language as simple and lucid as possible. Simple and accurate diagrams accompany the text. In the interest of the students efforts have been made to cover the syllabi of nearly all Indian Universities.

The book features both a text and a laboratory guide. It deals with the structures and physiological phenomenon of each system. General topics are also dealt in details. It is hoped that this book will not only meet the requirement of undergraduate students but also be useful as a guideline to the teachers in their teaching. It is also hoped that the general public will find that the book is not riddled with too much of scientific and technical jargon.

It is difficult to acknowledge adequately the assistance of all who have contributed to the preparation of this book. First of all the authors express their thanks to their friends and colleagues whose continuous inspiration have initiated them to bring out the present volume.

The authors are grateful to Dr. J. Bahadur, Professor and Head, School of Studies in Zoology, Jiwaji University Gwalior, Dr. A.K. Pandey, Professor and Head, School of Studied in Zoology, Vikaram University, Ujjain for their valuable suggestions.

Constructure criticism from students and teachers will be highly appreciated by the authors.

H.S. Bhamrah
Kavita Juneja

CONTENTS

1

ARTHROPODA

Characters and classification

Phylum Arthropoda (Gr. *arthros*= joint; *pods*=foot) contains the great majority of the known animals, about one million species, and many of them are enormously abundant as individuals. It includes such common and well-known forms as the crabs, shrimps, insects,spiders, scorpions. ticks, centipedes, as well as a host of other less familiar forms. *Cuvier* (1800) placed all these animals together with Annelida under the group *Articulate*. *Van Siebold* (1845) separated Annelida from Articulata. Annelida were placed under class Vermes while the Crustacea, Arachnida and Insecta under an independent group, the Arthropoda.

General Characters

1. Arthropoda are triploblastic,bilaterally symmetrical,metamerically segmented animals.
2. Body is covered with a thick chitinous cuticle forming an exoskeleton.
3. Body segments usually bear paired lateral and jointed appendages.
4. Musculature is not continuous but comprises separate striped muscles.
5. Body cavity is haemocoel. The true coelom is reduced to the spaces of the organs.

6. Digestive system complete; mouthparts adapted for various modes of feeding.

7. Circulatory system open with a dorsal heart, arteries and blood sinuses.

8. Respiration by general body surface, gills, tracheae or book-lungs.

9. Excretory organs are green glands or Malpighian trbules.

10. Nervous system with a dorsal nerve ring and a double ventral nerve cord.

11. Cilia are entirely absent from all parts of the body.

12. Sexes are generally separate and sexual dimorphism is often exhibited by several forms.

13. Fertilization is internal. Development is usually indirect through larval stages.

14. Parental care is also often will marked in many arthropods.

Phylum Arthropoda is divided into following classes:

class I	-	Crustacea
class II	-	Myripoda
class III	-	Insecta
class IV	-	Arachnida

CLASS - I. CRUSTACEA

1. The body is divisible into head, thorax and abdomen. when head and thorax are fused they form cephalothorax.

2. The exoskeleton is chitinous.

3. The body is metamerically segmented, bilaterally symmetrical and triploblastic.

4. The larval median eye is replaced by stalked compound eye of adult.

5. Each segment bears paired and jointed appendages.

6. The body cavity is filied with haemolymph and known as haemocoel.

7. Respiration takes place either by the general body surface or by gills.
8. Excretion occurs by modified coelomoducts, which may take the form either of maxillary glands or antennary (green) glands.
9. These are unisexual except many Cirripedia and some Isopods.
10. The development is indirect.

Class Crustacea is divided into 7 subclasses

Subclass - 1.Branchiopoda

1. Body has varying number of segments.
2. Antennules are small, uniramous and usually unsegmented.
3. Compound eyes are present.
4. Thorax bear four or more flat, leaf-like respiratory appendages also help in frilter feeding and locomotion.
5. Abdomen without appendages.
6. Parthenogenesis is common.

Subclass Branchiopoda is divided in three orders :

Order (i) Anostraca

1. The carapace is absent.
2. Eyes are stalked.
3. Antennal are uniramous, prehensile in male reduced in female.
4. Trunk elongated with 20 or more segments. Anterior Segment bear appendages.
5. The caudal styles are unjointed.

Example: *Branchipus, Streptocephalus, Artemia.*

Order (ii) Notostraca

1. A large dorsal shield- shaped carapace is present.
2. Eye are sessile i.e.stalkless.

3. Anteenae are reduced.
4. Another half of the trunk bears 35-70 pairs of appendages.
5. The caudal styles are many jointed.

Example : *Apus, Lapidurus.*

Order (ii) Conchostraca

1. The carapace is present but divided in two portions or valves.
2. Antennae biramous and large.
3. Eyes are sessile.
4. Caudal styles are in the form of unjointed, curved claws.

Examples : *Limnetis, Estheria,*

Order (iv) Cladocera

1. The carapaca bivalved,encloses the trunk but not the head and usually ends posteriorly in an apical spine.
2. Eyes are unstalked and united together.
3. Antennae biramous used for swimming.
4. Trunk segments 10-32 , each with a pair of appendages.
5. Caudal styles are unjointed claw-like

Example : *Daphnia.*

Subclass- 2. Ostracoda

1. Compound eye may or may not be present.
2. Carapace is bivalved and compressed which encloses entire body.
3. Head forms about half of the body.
4. Thorax usually bears only two pairs of cylindrical appendages.
5. Abdomen is reduced, without appendages and ends in a pair of caudal styles.
6. Antennules and antennae locomotory

7. Manibles have a palp.
8. Larva is Nauplius.
9. Parthenogenesis is common.

Subclass Ostracoda is divided into following orders:

Order (i) Myodocopa

1. Marine Crustaceans.
2. Carapace with antennal notches.
3. Antennae biramous.
4. Trunk appendages two pair.

Examples : *Cypridina, Concoecia.*

Order (ii) Cladocopa

1. Marine forms.
2. Carapace un-notched
3. Both pairs of antennae used in swimming.
4. Second pair of antennae with 2 branches

Example : *Polycope*

Order (iii) Podocopa

1. Maine or freshwater
2. Carapace Unnotched.
3. Second antennae leg - like, clawed at tips.
4. Trunk appendages 2 pairs.

Examples : *Cypris, Entocythere, Cythereis*

Order (iv) Platycopa

1. Marine forms
2. Carapace unnotched
3. Both pairs of antennae large but not used for swimming, second pair flatenned.

4. Trunk appendages single pair.

Example : *Cytherella.*

Subclass 3. Mystacocardia

1. Primitive marine forms.
2. Size microscopic, cylindrical.
3. Antennules and antennae long, prominent.
4. 4 eyespots, compound eyes absent.
5. Thorax 4- segments, each with one pair of simple appendages.
6. Excretion by both antennal and maxillary glands.
7. Ventral nerve cord paired.
8. Sexes separate, earliest larva is metanauplius.

Example : *Derocheilocarus*

Subclass - 4 . Copepoda

1. Fresh water or marine, free-living, commersal or parasitic.
2. Body small, elongated, composed of head, thorax and abdomen,
3. Typically 10 free trunk somites, with 4 lacking appendages. Appendages mostly biramous.
4. Three ocelli often fused as median eye, no compound eyes.
5. Antennules and antennae well developed and used for swimming.
6. Abdomen without appendages, ends in a pair of caudal styles.
7. Excretion by maxillary gland.
8. Eggs on abdomen of female in 1 or 2 egg sacs.

Subclass Copepoda is divided into 7 orders.

Examples : *Cyclops, Salmincola, Caligus,Doropygus, Calanus, Monstrilla.*

Subclass. 5. Branchiura

1. Ectoparasites of fresh water and marine fishes and some amphibians .
2. Body flat, carapace large, disk-like, covering head and thorax.
3. Abdomen small, bilked.
4. First maxillae modified as suckers.
5. Compound eyes present, sessile.
6. Thoracic appendages 4 pairs, large, biramous,some of them are provided with flagella.

Example : *Argulus*

Subclass. 6. Cirripedia

1. Marine forms.
2. Adults sessile, attached or parasitic; attached by cement gland on first antenna.
3. Carapace becomes mantle surrounding body.
4. Six or fewer pairs of slender and bristly biramous appendages behind mouth used in food gathering.
5. Abdomen vestigial, without appendages, ending in a pair of caudal styles.
6. Antennae and compound eyes disappear in adult.
7. Monoecious. Development includes free swimming nauplius which passes through a cypris stage which is enclosed in bivalve shell;

The subclass is divided into following orders.

Order (i) Thoracica

1. Non -parastic sedentary marine barnacles.
2. Body surrounded by calcareous shell.
3. Six pairs of biramous thoracic limbs.
4. Alimentary canal is present.
5. Hermaphrodite.

Example: *Balanus, Lepas.*

Order (ii) Acrothoracica.

1. Parasitic, boring in the shells of mollusc.
2. Body is covered by large mantle. Calcareous shell is absent.
3. Less than 6 pairs of thoracic appendages.
4. Alimentary canel is present.
5. Chitinous disc is present for attachment.

Example : *Alcippe*

Order (iii) Ascothoracica

1. Parasitizing corals and echinoderms.
2. Mouth appendages are modified into piercing and sucking organs.
3. Digestive tract with branches into mantle.
4. 6 pairs of trunk appendages.

Examples: *Laura, Petrarca.*

1. Parasitic in mantle cavity of *Alepas.*
2. Body without mantle and thoracic appendages.
3. Body maggot-like.
4. Anus absent.
5. Hermaphrodite.

Example : *Proteolepas.*

Order (v) Rhizocephala

1. Parasitic on crabs.
2. A mantle but no shell.
3. Appendages, gut are absent.
4. Body sac-like, with absorptive roots penetrating host.

Example : *Sacculina.*

Subclass-7- Malacostraca.

1. Mostly large sized crustaceans, marine and fresh water.
2. Body typically of 19 segments (5 head, 8 thorax and 6 abdomen).
3. Exoskeleton of head united with few or more thoracic segments to form cephalothoracic carapace.
4. Stalked compound eyes.
5. Abdomanal caudal style absent.
6. Development through zoaea stage. Nauplius stage rarely occurs.

Series [A]. Phyllocardia.

1. Abdomen of eight somites.

Order (i) Leptostraca.

1. Marine forms.
2. Body made up of 21 segments. Abdomen of 7 segments.
3. Thoracic appendages similar and foliaceous.
4. Telson has a pair of caudal styles.
5. Carapace bivaled. Eyes stalked.

Example : *Nebalia.*

Series [B]. Eumalacostraca

1. Abdomen of 6 or fewer somites.

Super order [a] Syncarida

1. Carapace is absent.

Order (i) Anaspidacea.

1. Fresh waters of Australia.

Example : Anaspides.

Super order [b] Peracardida

1. Carapace, when present, leaving 4 or more thoracic somites distinct.
2. Females with thoracic brood pouch where young develop.
3. Development direct without larval stage.

Order (i) Mysidacea

1. Marine forms.
2. Carapace covers almost entire thorax.
3. Eyes stalked.
4. Thoracic appendages are biramous.
5. Uropods form broad fan-like tail fin.

Example: *Mysis.*

Order (ii) Cumacea

1. Small, mostly marine, burrowing in surface sand or mud.
2. Carapace with 2 anterior extensions often joined over head.
3. Head and thorax greatly enlarged.
4. Abdomen slender, mobile.
5. Uropods slender.

Example : *Deastyles, Cumopsis.*

Order (iii) Tanaidacea

1. Minute, marine, upto depths of 3650 meters in mud or tubes.
2. Carapace is small and fused to first 2 thoracic segments,
3. Second thoracic appendage with chela.
4. Telson unjointed fused with last abdominal segment.
5. Uropods slender.

Examples : *Tanais, Apseudes.*

Order(iv) Isopoda.

1. Mostly marine, some freshwater and few parasitic.
2. Body usually depressed dorsoventrally.
3. Carapace is absent. Abdomen short, partly or all fused.
4. Gills and heart abdominal.

Examples : *Ligia, Asellus, Hemioniscus.*

Order (v) Amphipoda.

1. Mostly marine, some freshwater and terrestrial.
2. Body often laterally compressed.
3. Carapace is absent.
4. Telson usually distinct.
5. Gills and heart thoracic.

Example : *Gammarus, Caprella, Cyamus.*

Super -order [c] Hoplocarda

1. Head with two movable anterior segment bearing eyes and antennules.

Order (i) Stomatopoda

1. Marine on bottom in sand or cervices.
2. Antennal scale enlarged.
3. Gills on abdominal appendages modified for raptorial feeding
4. Carapace small.

Examples : *Squilla , Pseudosquilla, Chloridella*

Super order[d] Eucardia

1. Carapace is present, fused to and covering all the thorax.
2. Eyes stalked.
3. Gills thoracic.

4. Brood pouch absent.
5. Development usually indirect with a zoaea larva.

Order (i) Euphausiacea

1. Marine, small, pelagic forms.
2. Carapace does not cover sides of body and the gills.
3. Thoracic appendages all biramous and similar, none modified as maxillipeds.
4. Pleoplds have long setae, used for swimming.
5. Larva is nauplius.

Examples : *Euphausia, Thysanopoda.*

Order (ii) Decapoda

1. Mostly marine forms.
2. Carapace covers the entire thorax.
3. Thoracic appendages are modifies as 5 pairs of walking legs and three pairs of maxillipedes.
4. Thoracic appendages mostly uniramous.
5. Statocyst present.
6. Gills usually present in three series on the thorax.

Decapoda is divided into 3 suborders:

Sub order (a)Macrura

1. Abdomen is well developed and elongated.
2. The abdomen terminates in an expanded fan-like tail fin composed of telson and the uropods.
3. Eyes are not enclosed in orbits.

Examples: *Lucifer, Palaemon , Homarus, Astacus, Cambarus.*

Suborder(b) Anomura

1. Abdomen is more or less reduced and flexed.

2. Tail fin is also reduced.

Examples : *Eupagurus, Hippa*

Suborder (c) Brachyura

1. Abdomen greatly reduced , hard, permanently flexed under cephalothorax.
2. Antennules and the eyes are both capable of being retracted into cavities.
3. Rostrum and uropods absent.
4. First legs usually large and chelate.

Examples : *Cancer, Inachus, Uca, Libinia, Inachus*

CLASS - II MYRIPODA

1. Terrestrial and air breathing arthropods.
2. Body is usually elongated with numerous segments,each bearing one or two pair of appendages.
3. Head is distinct, bearing a pair of antennae, mandibles, eyes and maxillae.
4. Respiration by tracheae.
5. Excretoty organs are one or two pairs of Malpighian tubules associated with hindgut.
6. Sexes are separate.

Subclass- 1 Progeneata

1. The genital apertures are situated near the anterior end of the body.

Order (i) Pauropoda.

1. The trunk segments are eleven and anal segment is also present.
2. There are 9 pairs of legs.
3. The head bears one pair of antennae, one pair of mandibles and one pair of maxillae.

4. These are without blood vascular system and trachea.

Example : *Pauropus.*

Order (ii) Diplopoda

1. Body is usually cylindrical.
2. Poision bears numerous segments each with two pairs of legs (expect first four).
3. Genital opening on the third segment behind the head.

Example : *Julus, Millipede.*

Order (iii) Symphyla

1. There are not more than 12 segments in the trunk.
2. The head bears one pair of antennae, one pair of mandibles and 2 pairs of maxillae.
3. There is a single pair of tracheae.

Example : *Scolopendrella, Scutigerella,*

Subclass - 2. Opisthogoneata

1. Genital apertures are situated at the posterior extremity of the body.

Order (i) Chilopoda

1. Body is usually dorsoventrally flatenned.
2. Head bears many jointed antennae, a pair of mandibles and two pairs of maxillae.
3. Poison jaws are present.
4. Trunk bears 15 -117 segments each with a single pair of legs.

Examples : *Scolopendra, Scutigers,*

CLASS III INSECTA

1. These are air breathing, mostly tarrestrial and rarely aquatic arthropods.
2. Body is divisible into head, thorax and abdomen.
3. Head bears a pair of antennae, compound eyes, mandibles and two pairs of maxillae.
4. Thorax cosists of three segments bearing 3 pairs of legs and 2 pairs of wings.
5. Abdomen 7-11 segments which are without appendages in adult.
6. Liver absent but salivary glands are present.
7. Respiration by tracheae.
8. Excretion by Malpighian tubules.
9. Sexes separate.
10. Development is usually inderect, completed by complicated metamorphosis. Sometimes direct without larval stage.

Class Insecta is divided into two subclasses.

Subclass I Apterygota

Subclass II Pterygota.

Subclass - I- Apterygota.

1. Wingless, primitive insects.
2. Metamorphosis is absent.

Order (i) Protura

1. Minute, elongate, whitish insects.
2. Antennae, eyes and ocelli are lacking.
3. Wings are absent.
4. Abdomen short consists of 12 segments. Bilateral styli on first three segments.
5. Tracheae lacking.

6. Malpiglian tubules are small papillae.

Example : *Acerentulus*

Order (ii) Collembola.

1. Minute, somewhat tubular or globose forms.
2. Eye patches consist of 1 to several lateral ocelli, dorsal ocelli lacking. Antennae short. usually 4- segmented.
3. Abdomen consists of 6 segments, lobelike organ on first segment. 4th segment bears furcula.
4. Mouth parts entognathus, chewing type.

Examples : *Podura, Isotoma, Sminthurus.*

Order (iii) Archeognatha

1. Small, elongate, many with scales on part of body.
2. Mouth parts are ectognatous chewing type.
3. Antennae filiform, large compound eyes and 3 ocelli are present.
4. Abdomen small bilateral styli on venter of several segments.
5. Females with elongated jointed ovipositor.

Example : *Machilis*

Order (iv) Thysanura.

1. Small, elongated insects. Usuelhy covered with scales.
2. Compound eyes usually present. O - 3 ocelli.
3. Abdomen small, betateral style on venter of several segments.
4. Abdomen consits of 10 or 11 segments.

Examples : *Lepisma, Japyx.*

Subclass II Pterygota

1. Adults are winges, some are wing less.
2. metamorphosis complete or incomplete.

Division (A) Exopterygota (Heterometabola).

1. Wings develop externally as buds.
2. Metamorphosis simple without pupal stage.

Order (i) Odonata

1. Medium to very large sized insects.
2. Head large, mobile bearing short antennae, large compound eyes and 3 dorsal ocelli.
3. Two pairs of membraneous wings, venation net-like.
4. Abdomen elongate. Adult meles with gonopores on 9 th segment and complex penis on second segment.
5. Metamorphosis incomplete. The nymphe aquatic, predatory with caual tracheal gills.

Examples : Dragon flies (*Lestes*, Damsel flies.

Order (ii) Ephemeroptera

1. Soft bodied, fragile, small to medirm sized.
2. Mouth parts chewing type in nymph and vestigial in adults.
3. Antennae short. Compound eyes and 3 ocelli are present.
4. Two pairs of membranous wings. Forewings larger than hind wings.
5. Abdomen with a pair of long cerci and often with a median caudal filament.
6. Metamorphoses incomplete. Nymph bears gills of the first 4 - 7 segments.

Example : Mayflies (*Ephemera*).

Order (iii) Placoptera

1. Small to medium sized, soft- bodies, elongate, flatenned insects.
2. Mouth parts chewing type, usually vestigial in adults.

3. Antennae long and tapering. Well developed compound eyes, usually 3 dorsal ocelli are present.
4. Two pairs of membranous wings, males of some species apterous or brachypterous.
5. Metamorphosis incomplete. Nymphs aquatic with tracheal gills.

Examples : Stoneflies (*Isoperla* , *Pteronarcys*)

Order (iv) Embioptera

1. Minute to small, elongated insects.
2. Head large bearing fileform antennae, compound eyes present, ocelli absent.
3. Mouth parts chewing type.
4 Thorax nearly as long as abdomen.
5. Some males with 2 pairs of membranous wings of nearly equal size and shape. Females and some males winless.
6. Metamorphosis gradual in male but absent in female.

Example : Web -spinners (*Oligotoma*).

Order (v) Phasmida

1. Large sized insects. Body elongate, cylindrical with short prothorax,large meso and metathorax.
2. Head broad and bears a pair of long filiform or moniliform antennae. Compound eyes are present. 2-3 or no ocelli.
3. Mouth parts biting type.
4. Wings presint or absent.
5. Abdomen 11 segmented. Ovipositor small and somewhat concealed. Cerci short, unsegmented.

Examples: *Phyllium, Carausius* (stick insect).

Order (vi) Orthoptera

1. Minute to very large sized insects.

2. Mouth parts chewing type. All gradation between hypognathous and opisthognathous.
3. A pair of antennae of variable structure.
4. Compound eyes and usually 2 or 3 ocelli are present.
5. Two pairs of wings are present. Forewings hard, narrow, parchment like opaque with indistinct venation forming tegmine (wing covers).
6. Abdomen 11 segmented. Females in several species with well developed ovipositors. Cerci variable, short, long, clasperlike, segmented or unsegmented.

Examples : Grasshoppers (*Poecillocerus*), Locusts (*Schistocerca*), Crickets (*Gryllus*), Mole cricket (*Gryllotalpa*)

Order (v) Dermaptera

1. Small to medium sized insects.
2. Mouth parts chewing type.
3. Antennae long slender. Compound eyes present in most of the forms may be vestigial or absent in some forms.
4. Fore wings short, leathery and veinless,hind wings membraneous, broad. fan-like
5. Abdomen11 segmented bearing a pair of forceps-like cerci at the end.
6. Metamorphosis gradual.

Example : Earwigs (*Forficula*)

Order (vii) Grylloblattoidea

1. Medium to large, some what elongate and slender.
2. Mouthparts parts biting and chewing type.
3. Antennae filiform. compound eyes reduced or absent. Ocelli also absent.
4. Wings absent.
5. Abdomen well developed with sword shaped ovipositor. Cerci long.

6. Metamorphosis simple.

Example : *Grylloblatta.*

Order (viii) Isoptera

1. Social and polymorphic insects living in colonies under a caste system.
2. Mouth parts chewing type.
3. Compound eyes are present in all winged forme, present or absent in apterous forms. O or 2 dorsal ocelli are present.
4. Antennae short, moniliform or filiform.
5. Wings when present, two pairs, similar in size, shape and venation.
6. Cerci short and simple.
7. Metamorphosis simple.

Examples : Termites (*Zootermopsis Nasutitermes*).

Order (ix)Blattaria.

1. Medium to large sized insects, some what dorsoventrally flatenned.
2. Mouth parts chewing type.
3. Compound eyes well developed, 2 ocelli in some. Antennae long filiform.
4. Wings present or absent, when present forewings tegmina, hind wing membranous.
5. Pair of styli on 9th sternum of males.Ovipositor reduced and concealed by 7th abdominal sternum.
6. Metamorphosis simple.

Example; *Periplaneta, Blattella.*

Order (x) Mantodea

1. Medium to large sized insects. Body usually elongate, somewhat cylindrical, with elongate prothorax.

2. Head small, triangular and bears large compound eyes with 3 or more ocelli.
3. Mouth parts chewing type.
4. Raptorial forelegs at anerrior end of elongated prothorax.
5. Wings usually present in males, reduced or absent in females.
6. Abdomen comparatively short, multisegmented cerci, pair of styli usually associated with 9th sternum in males.
7. Metamosphosis simple.

Example : Praying mantids (*Tenodera*).

Order (xi) Zoraptera

1. Minute sized insects.
2. Mouth parts chewing type.
3. Compound eyes and ocelli present in winged forms but absent in wingless forms.
4. Wings present or absent when present forewings larger than hind-wings.
5. Abdomen short one segmented cerci.
6. Metamorphosis simple.

Example: *Zorotypus*.

Order (xii) Psocoptera

1. Minute sized insects . Head capsule large compared to rest of body.
2. Mouth parts chewing type. Labial silk-glands are present.
3. Compound eyes weakly or strongly developed. Ocelli 3 in winged forms absent in wingless forms.
4. Wings when present, two pairs, menbranous.
5. Cerci absent.
6. Metamorphosis simple.

7. Examples : Book - lice (*Liposcelis*), *Peripsocus*.

Order (xiii) Mallophaga

1. Small, apterous, ectoparasitic insects.
2. Body is broad or elongated, flatenned dorsally.
3. Head generally large. Compound eyes reduced. Ocelli absent.
4. Mouth parts biting type.
5. Cerci absent.
6. Metamorphosis absent.

Examples : Chewing lice or biting lice (*Trichodectes, Menacanthus*).

Order (xiv) Anoplura.

1. Minute, dorsoventrally flatenned, ectoparasitic.
2. Head narrower than thorax.
3. Mouth parts piercing and sucking, retracted into head when not feeding.
4. Compound eyes weakly developed or lacking,ocelli also absent.
5. Antennae short 3-5 segments, setiform.
6. Wings are absent.
7. Legs adapted for clinging to the hairs of the host and each terminating in a curved claw.
8. Metamorphosis absent.

Examples: Human body - louse (*Pediculus*), *Phthirus*.

Order (xv) Thysanoptera.

1. Minute to small sized insects, Body slender.
2. Mouth parts rasping and sucking type: Asymmetrical, the right mandible reduced or absent.
3. Small, compound eyes; 3 dorsal ocelli in winged forms absent in wingless forms.

4. Antennae short 6 - 10 segments.
5. Wings two pairs, narrow fringed; coupted by basel hooks.
6. Abdomen 10 -11 segments with an ovipositor.Cerci lacking.
7. Metamorphosis simple.

Example : Thrips (*Heliothrips*).

Order (xvi) Hemiptera

1. Medium to very large sized insects.
2. Mouth parts piercing and sucking type.
3. Antennae 2 - 10 or rarely 25 segmented.
4. Eyes large with or without ocelli.
5. Wings when present, two pairs, fore wings usually thickened basally and membranous apicalhy (suborder Heteroptera) and wholly membranous (suborder Hereroptera) and wholly membranous (suborder Homoptera).
6. A triangular plate is located between the bases of the wings.
7. Anal respiratory filaments are present in some aquatic forms and anal filaments in male coccids.
8. Metamorphosis gradual.

Examples : Giant water bug (*Belostoma*), Bedbug (*Cimex*), water strider (*Gerris*), Squash bug (*Anasa*), Leaf hopper (*Paraphlepsius*), Cicada (*Magicicada*).

Division (B) Endopterygota (Holometabola)

1. Wings develop internally.
2. Metamorphosis complete.

Order (i) Coleoptera

1. Minute to large sized insects.
2. Antennae variable in shape and size usually 11 segmented.

3. Compound eyes present or absent. Ocelli usually absent but present in some groups. Larva without compound eyes but lateral orelli present.

4. Forewings sclerotized elytra which protect the membranous hindwings at rest.

5. Legs typically cursorial,some adapted for swimming , jumping or digging.

6. Abdomen usually 10 segmented . Cerci absent.

7. Mouth parts biting and chewing types.

8. Larvae diverse usually catterpillar - like or maggot-like.

9. Metamorphosis complex.

Examples : Ground beetle (*Calosoma*), Blister beetle (*Epicauta*), Alfalfa beetle (*Hypnera*).

Order (ii) Strepsiptera.

1. Minute sized mostly endoparasitic insects.

2. Mouthparts with reduced mandibles.

3. Compound eyes present in males and free - living females.

4. Antennae flabellate in males,absent in females.

5. Wings two pairs in males, forewings clublike, hindwings membranous and fan - like.

6. 3 - 5 genital pores in the abdomen of femals

Example : *Stylops*.

Order (iii) Neuroptera

1. Small to very large-sized , soft - bodied insects.

2. Mouth parts chewing type in adult and grasping - sucking type in larvae.

3. Compound eyes present. Ocelli present or absent.

4. Antennae filiform.

5. Wings two pairs, similar in size and appearence.

6. Cerci absent. Variously modified ovipositors are present.
7. Metamorphosis complete/

Examples : Green lacewing (*Chrysopa*), Snakefly (*Agulla*), Dobsonfly (*Corydalis*).

Order (iv) Mecoptera

1. Small to medium sized, slender, predaceous insects.
2. Head usually elongated. Mouth parts chewing type.
3. Compound eyes are present. ocelli 0-3. Antennae long and filiform.
4. Wings 2 pairs, membranous, commonly with dark spots.
5. Cerci shorts, simple or two segmented.
6. Metamorphosis complete.

Examples : Scorpionfly (*Panorpa*), Hanging scorpionfly (*Bittacus*).

Order (iv) Diptera

1 Minute to medium sized, diurnal, noctuirnal, terrestrial or aquatic insects.
2. Mouth parts variously modified for sucking, lapping, piercing and sponging.
3. Compound eyes present. Most species with 3 ocelli.
4. Wings are present, fore wings developed for flight while hind wings are reduced to halters.
5. Cerci present or absent.
6. Metamorphosis complete.

Examples; Mosquitoes (*Aedes, Anopheles, Culex*), Black fly (*Simulium*), House fly (*Musca*),Fruit fly (*Drosophila*).

Order (v) Siphonoptera .

1. Minute, hard - bodied bilaterally compressed insects.
2. Mouth parts piercing and sucking type. In larvae chewing type.

3. Compound eyes lacking. Generalhy 2 ocelli.
4. Antennae short contained within antennae grooves.
5. Wings absent.
6. Male terminalia highly modified with one pair of 2 segmented claspers.Cerci lacking.
7. Metamorphosis complete.

Examples : Human fleas (*Pulex*) Polutary flea (*Echinophaga*),Rat flea (*Xenopsylla*), Dog and cat flea (*Ctenocephalides*).

Order (vi) Trichoptera

1. Small to medium sized insects.
2. Monthparts are nonfeeding in adult stage. Adapted for inbibing fluid. Mandibles weakly developed or lacking. Larval with chewing type.
3. Compound eyes present, O or 3 ocelli.
4. Antennae range from setaceous to filiform.
5. Wings 2 pairs, membranous, hind wings broader then forewings; covered with modified hair - like setae.
6. Larvae aquatic.
7. Cerci one or two segmented.
8. Metamorphosis holometabolic.

Example : Caddis fly (*Philopotamus*).

Order (vii) Lepidoptera

1. Medium to large sized insects .
2. Body and wings are completely corered by flat overlapping scales and hairs.
3. Mouth parts usually long, sucking proboscis.
4. Large compound eyes; O or 2 dorsal ocelli.
5. Antennae always prominent, vary in form.

6. Wings two pairs, forewings often largest.
7. Metamorphowis complete.
8. Larvae called caterpillars with 3 pairs of thoracic legs and 2 -4 pairs of abdominal prolegs.

Examples : Butterflies (*Danaus*), Moths (*Bombyx*).

Order (viii) Hymenoptera.

1. Minute to large sized , social or parasitic insects.
2. Antennae 12 segmented in male and 13 segmented in females.
3. Mouth parts variable, chewing to lapping or sucking.
4. Compound eyes usually well developed. Commonly three dorsal ocelli.
5. Wings two pairs of membranous . Hindwings smeller then forewings, many wingless.
6. Abdomen with basal pedical.
7. Female with conspicuous ovipositor modified into saw, drell or sting.
8. Metamorphosis complete.

Examples : Honey bees (*Apis*) , Thief ant (*Solenopsis*) Warp (*Vespa*), Ants etc.

CLASS IV ARACHNIDA

1. These are air breathing, mostly terrestrial arthropods.
2. Body is divisible into cephalothorax and abdomen .
3. Cephalothorax bears two pairs of jointed appendages the cheliceral and padipalpi and four pairs of legs.
4. Antennae and true jaws are absent.
5. Eyes are sessile and simple.
6. Resperation by tracheal or book lungs or book - gills.
7. Heart is dorsal and tubular.

8. Excretion by coxal glands or Malpighian tubules or both.
9. Sexes are separate with conspicuous sexual dimorphosim.
10. Development direct.

Order (i) Eurypterida

1. Extinct forms.
2. Exo- skeleton is characteristically sculptured.
3. Cephalothorax is relatively small and the abdomen bears 12 free segments and a telson.
4. Cephalothorax bears 6 pairs of appendages.

Example : *Eurypterus.*

Order (ii) Xiphosura

1. Marine arachnids burrowing in the sand.
2. These animals have large subcircular, apparently unsegmented body with two body divisions - anterior large cephalothorac (prosoma) and pasterior small abdomen (opisthosoma), the two - joined by a hinge.
3. The cephalothorax bears a pair of short chelate appendages and 5 pairs of legs.
4. Abdomen bears operculum.
5. Respiration takes place by lamelliform gills attached to the abdominal appendages.
6. Sexes separate. Development includes a tritobite larva.

Example : King crab (*Limulus*)

Order (iii)Scorpionidea.

1. Terrestrial animals found under stones in tropical and subtropical regions.
2. Body is divisible into cephalothorax and abdomen.
3. Prosoma bears a pair of chelicerae, a pair of pedipalpi and four pairs of walking legs.

4. Respiration takes place by 4 pairs of book -lungs.

Examples : *Palamnaeus, Buthus.*

Order (iv) Pedipalpida

1. Body consists of cephalothorax and abdomen.
2. The chelicerae are simple and non-chelate.
3. Respiratory organs are two pairs of book lungs

Example: *Phrynichus.*

Order (v) Araneida

1. Found under stones, bark, leaves etc.
2. Body consists of an undivided cephalothorax and an unsegmented abdomen.
3. Cephalothorax bears 6 pairs of appendages.
4. Chelecerae are subchelate with poison glands. Pedipalpi are simple and 6 jointed.
5. Eight eyes are arranged dorsally on the head.
6. Respiration by book- lungs or book - lung with tracheae.

Example; Spider (*Aranea*), *Argiope.*

Order (vi) Palpigradi.

1. Small arachnids.
2. Body consists of cephalothorax and an abdomen of 10 segments.
3. Chelecerae are chelate and pedipalpi leg-like.
4. Respiratory organs are 3 pairs of bood-lungs

Example : *Koenenia.*

Order - (vii) Solifugae

1. Body divisible into prosome and opisthosoma.
2. Prosome consists of three fused and last three segments.

3. Opisthosoma compsoed of 10 segments is not marked off from prosoma.
4. Chelecerae are large and chelate. Pedipalpi are elongated and leg-like.
5. Respiration by tracheae. Poison gland are absent.

Example : *Goleodes.*

Order (viii) Pseudoscorpionidea

1. Minute arachnids having the appearence of scorpion, found under the tree barks.
2. Body consists of cephalothorax and an abdomen of 12 segments.
3. A pair of spinning glands are present.
4. Respiration by tracheae

Examples: *Chelifer.*

Order (ix) Recinulei

1. Body consist5s of cephalothorax and an abdomen of 9 segments.
2. Cephalothorax and pedipalpi are *cucullus.*
4. Respiratory organs are tracheae.

Exanple : *Cryptocellus*

Order (x) Phalangida.

1. Body consists of unsegmented cephalothorax and an abdomen of 10 segments.
2. Cheleeerae are chelate and pedipalpi leg - like.
3. Spinning glands are absent.
4. Resperatory organs are tracheae.

Example : Obligolophus.

Order (xi) Acarina

1. Body is unsegmented.
2. Chelecerae and pedipalpi are usually small and associated with the mouth parts.
3. Mouth parts are piercing and sucking type.
4. Respiration by tracheae or skin.

Example : Ticks (*Lxodes*), Mites (*Chorioptes*)

PHYLUM ONYCHOPHORA

1. Aberrant forms as its several characters are of phylum Annelida and Arthropoda combind.
2. Body - soft, worm -like covered by papillae and hairy spines.
3. Cuticle is free, having transerve rings, metamerism not marked externally.
4. Jaws are only one pair.
5. Series of shout, fleshy , legs pairs provided with two claws.
6. Respiratory organs are tracheae.
7. Sexes are separate.
8. Regarded as living fossil.

Example : *Peripatus*.

BRANCHIPUS

Phylum	-	Arthropoda
Class	-	Crustacea
Subclass	-	Branchiopoda
Order	-	Arostraca
Genus	-	*Branchipus*

1. It lives in fresh water lakes, ponds etc.
2. It is commonly known as 'brine shrimp'
3. Body is elongated pinkish, semitransparent and differentiated into head, thorax and abdomen.

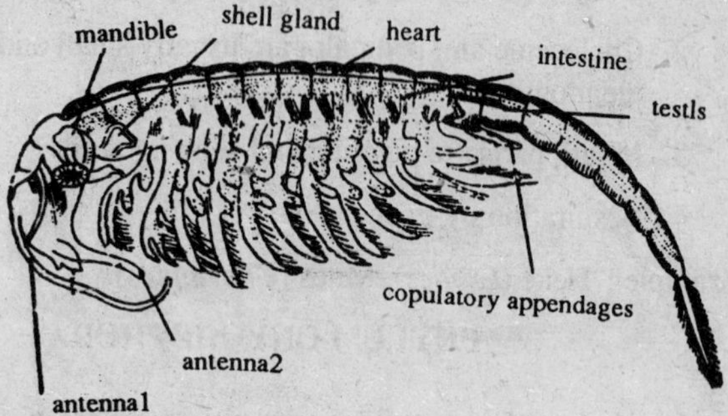

Fig. 1.1. *Branchipus*

4. Carapace is absent.
5. Eyes are situated on unjointed stalkes.
6. Antennae many and are modified into prehensile organs.
7. The cacedal styles are enjointed.
8. Eggs are laid in mud during summer and hatch in the next summer.

APUS

Phylum	-	Arthopoda
Class	-	Crustacea
Subclass	-	Brnachiopoda
Order	-	Notostraca
Genus	-	*Apus.*

1. It is a fresh water crustacean found in most parts of the world especially in U.S.A. Kansas, Texas, Mexico etc.
2. It is commonly called 'tadpole shrimp'
3. Body is elongated with broad shield-shaped carapace like horse zhoe.
4. Head is broad and depressed bears paired eyes, a median eyes and a dorsal organ above the antennules and antenna below.

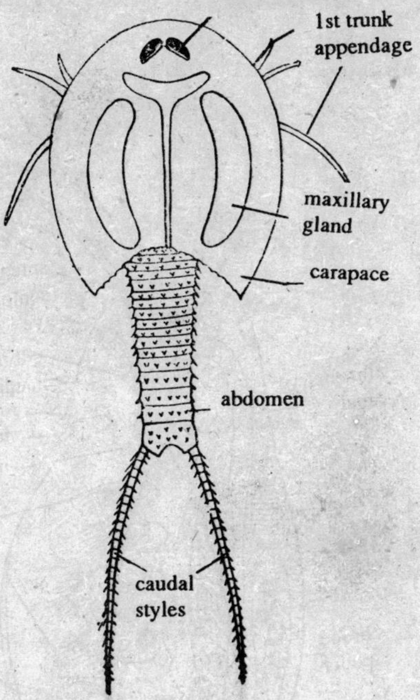

Fig. 1.2. *Apus*

5. Shell glands are present on the lateral surface of carapace.
6. The original segments of the body seem replaced by secondary annulation.
7. Caudal styles are long, filamentous and jointed.
8. Males are rare, hence reproduction commonly occurs by parthenogenesis.

DAPHNIA

Phylum	-	Arthropoda
Class	-	Crustacea
Subclass	-	Branchiopoda
Order	-	Cladcera
Genus	-	*Daphnia*

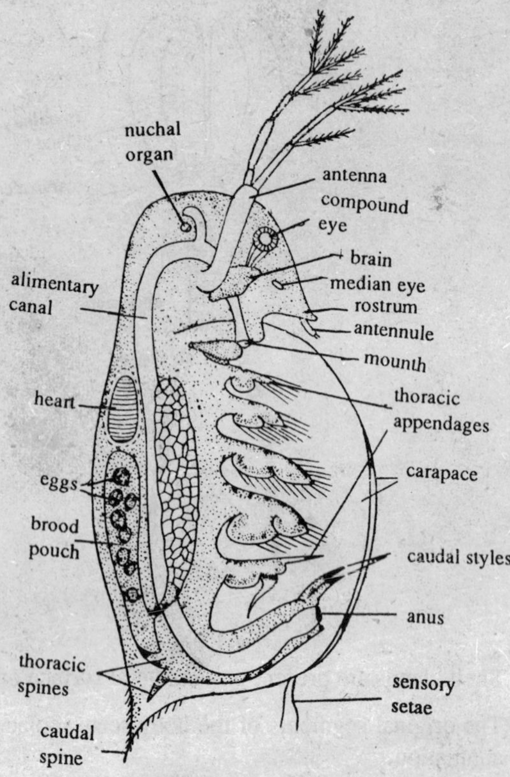

Fig. 1.3. *Daphnia*

1. It is fresh water crustacean found in ponds, pools and ditches.
2. It is commonly known as 'water flea'.
3. The body except head is covered by folded carapace.
4. Head and abdomen are turned downward.
5. Head bears sessile eyes, nuchal sense organs above it, a pair of entennules, a pair of biramous antennae, large mandibles and two pairs of maxillae.

6. Large biramous antennae are the cheif organ of locomotion.
7. Thorax bears usually five pairs of leaf like appendages.
8. Abdomen is devoid of appendages.
9. The female consists a of a brood pouch between posterior part of carapace and abdomen in which eggs are stored.
10. The alimentary canal is simple
11. Heart is a little sac.
12. It is a delicious food for many aquatic animals.
13. Sexes are separate. Female carries eggs and embryos in a large brod pouch situated between abdomen and posterior part of the carapace.
14. No larval stage occurs.

CYCLOPS

Phylum	-	Athropoda
Class	-	Crustacea
Subclass	-	Copepoda
Genus	-	*Cyclops*

1. *Cyclops* is found in fresh water ponds, ditches and also in brackish water.
2. The body is elongated, somewhat oval in shape.
3. The cephalothrax is formed by the fusion of head and the first thoracic segments.
4. Carapace is absent.
5. The cephalothorax is followed by the five free segments of the thorax and an abdomen of four segments.
6. The last segment of the thorax is fused in the female which first abdominal segment and bears the genital openings.
7. The antennules are large, entire and segmented and they are meant for copulation in male.
8. The antennae are short and the mandibles have a palp.

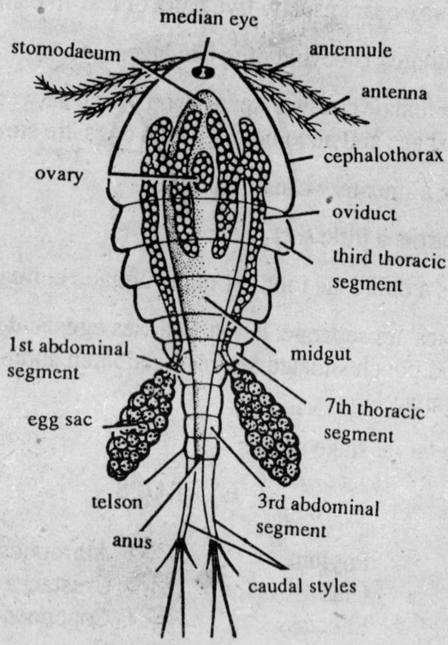

Fig. 1.4. *Cyclops*

9. There are two pairs of maxillae following with are maxillipedes and which are the first thoracic appendages. Rest are used for swimming.
10. Abdomen is devoid of appendayes except the caudal styles.
11. Sexes are separate. Mature female carries two egg sacs attached to the abdomen.

CYPRIS

Phylum	-	Arthropoda
Class	-	Crustacea
Subclass	-	Ostracoda
Genus	-	*Cypris*

1. It is free-swimming and occurs in fresh water stagnant ponds.
2. It has about 2 mm long, unsegmented, laterally compressed,

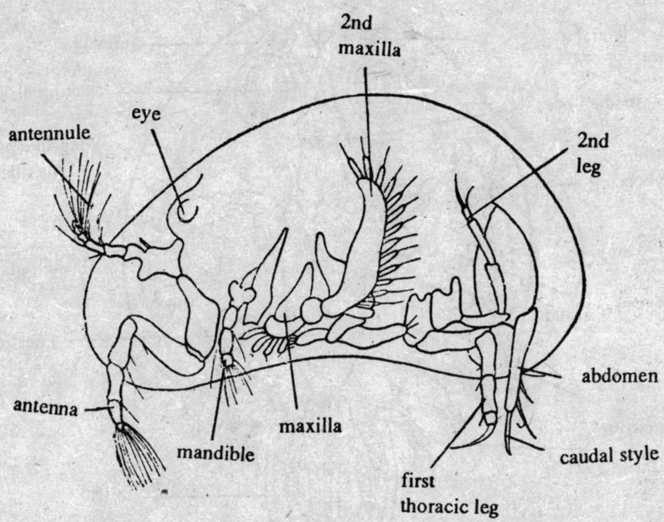

Fig. 1.5 *Cypris*

clam-like body enclosed in a bivalved shell. The valves of the shell are closed by a transverse adductor muscle.

3. Compound eyes are absent but there is a prominant median eye. Antennule and antennae are large.

4. It is an omnivorous fitter feeder.

5. Abdomen is devoid of appendages and terminates into two caudal styles.

6. Respiration by integument.

7. Development by parthenogenesis.

ARGULUS

Phylum	-	Arthopoda
Class	-	Crustacea
Subclass	-	Branchura.
Genus	-	*Argulus*

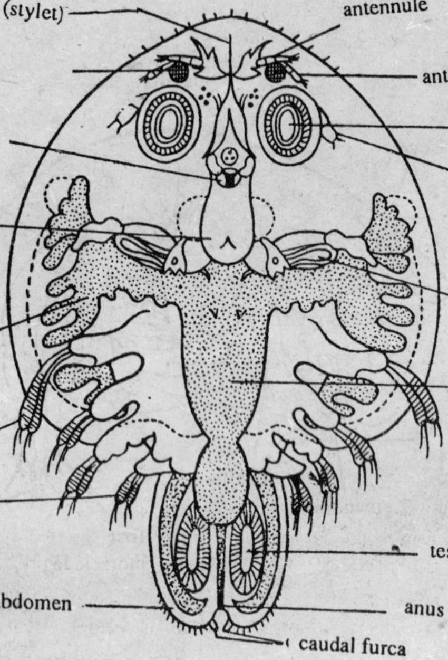

Fig. 1.6. *Argulus*.

1. It is an ectoparasite on fresh water fishes and usually found in the walls of branchial cavity.
2. It is commonly known as 'carp louse'.
3. Body consists of an oval flaterred cephalothorax and a small bilobed abdomen.
4. Mandibles and maxillae form piercing organs and are surrounded by a sucking tube called proboscis.
5. A poisonous spine, the stylet, lies in front of the proboscis.
6. Second maxillace are divided into two parts, an anterior sucking disc for attachment and a posterior leg like part.
7. A pair of large compound eyes and a median eye are present on the head.

8. Thorax bears 4 pairs of swimming appendags.
9. Gills are absent, respiration probably through the carapace surface.

LUCIFER

Phylum	-	Arthropoda
Class	-	Crustacea
Subclass	-	Malacostraca
Order	-	Decapoda
Genus	-	*Lucifer*

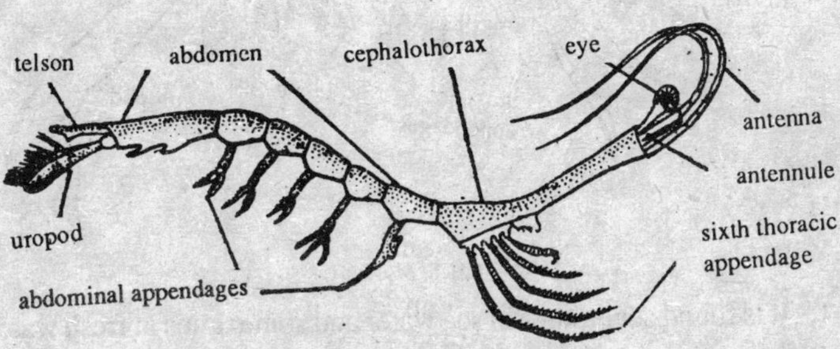

Fig. 1.7. *Lucifer*

1. It is a pelagic crustacean found in India, Asia and Europe.
2. The animal is minute, delicate and slender.
3. Body is divisible into cephalothorax and abdomen.
4. Cephalothorax is smaller than abdomen.
5. Head os elongated and bears long antennae and stalked eyes.
6. The thoracic appendages are non-chelate and the last two thoracic somites are without appendaues.
7. Gills are absent.
8. Development indirect through nauplius larva which moults through zoaea and a mysis stages.

MYSIS

Phylum	-	Arthropoda
Class	-	Crustacea
Subclass	-	Malacostraca
Order	-	Mysidacea
Genus	-	*Mysis*

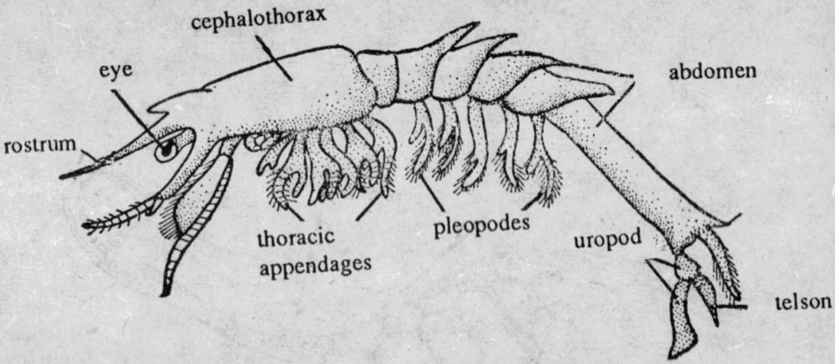

Fig. 1.8. *Mysis*

1. It is found soemtimes in sea water and some times in fresh water.
2. *Mysis* is a small transparent, shrimp-like form.
3. Body is bilaterally compressed and elongated measuring from 2-6mm in length.
4. Carapace covers the entire thorax except the last two segments.
5. Head bears antennules, antennae and a pair of stalked eyes.
6. First pair of thoracic appendages are modifies as maxillipedes and the rest are biramous serving as swimming organs.
7. Brood pouch is attached to the posterior thoracic segments.
8. Development takes place with the brood pouch, so there is no larval stage.

GAMMARUS

Phylum	-	Arthropoda
Class	-	Crustacea

Subclass	-	Malacostraca
Order	-	Amphipoda
Genus	-	*Gammarus*

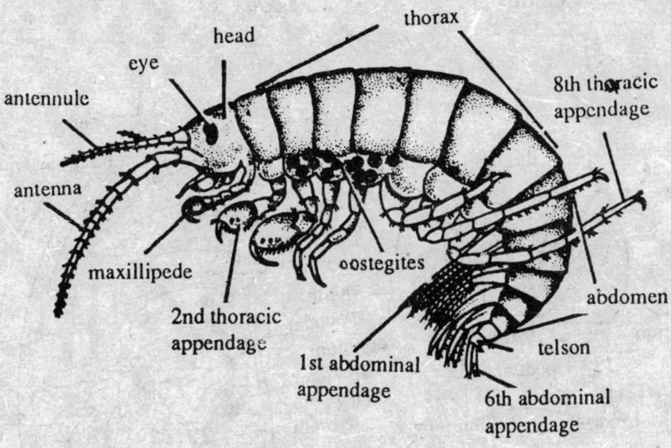

Fig. 1.9. *Grammarus*.

1. It is marine as well as found in fresh water. It is commonly called 'sand-flea.
2. Body elongated and laterally compressed.
3. Cephalothorax is formed by the fusion of head and first thoracic segment.
4. Carapace lacking.
5. Head bears a pair of antennule, antennae and sessile eyes.
6. First thoracic segment bears maxillipeds and the remaining seven segments bear walking legs. First two pairs of legs are chelate.
7. The abdomen is of a considerbale size and flexed ventrally, bears a bifid telson.
8. First three pairs of abdominal appendages are used for swimming and rest for jumping on the gound.

9. It feeds an dead and decaying organisms.

LEPAS

Phylum	-	Arthropoda
Class	-	Crustacea
Subclass	-	Cirripedia
Order	-	Thoracica
Genus	-	*Lepas*

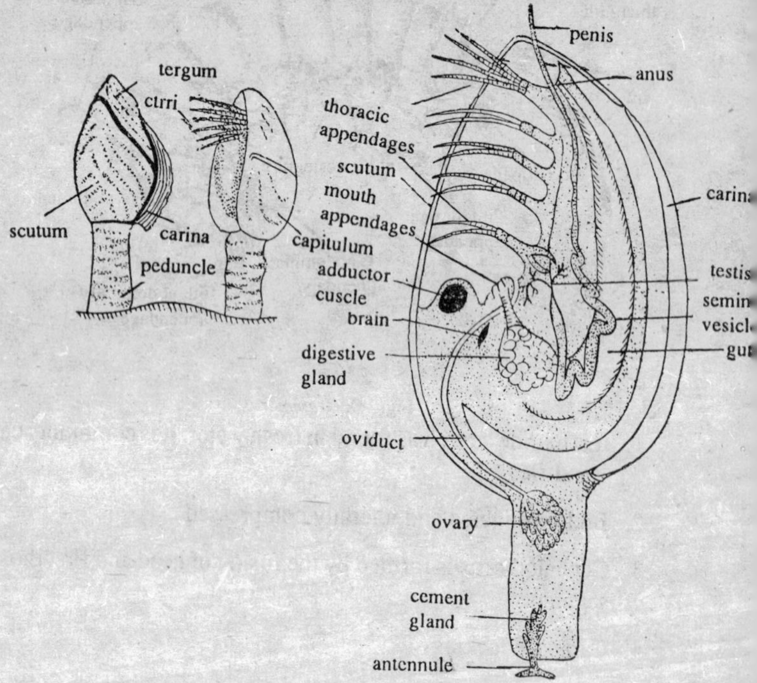

Fig. 1.10 *Lepas*

1. It is sessile in habit and found attached in the clusters of floating objects like logs and the bottom of ships in the sea.

2. It has cosmppolitan distribution.

3. It is commonly called the 'goose barnacle'.

4. Body consists of a long stalk or *Pedicel* and the body proper or *capitulum.*
5. Pedicle is covered by wrinkled skim.
6. The cepitulum is enclosed in a kivalve carapace strengthened by 5 calcareous plates, 2 proximal scuta, 2 distal terga and a single dorsal carina.
7. The pedicle is really the highly modified anteriormost part of the head contains the vestige of the antennules, ovaries and the cement gland which provide the stickly secretion for attachment.
8. Antennae and paired eyes are absent.
9. Mouth is provided with a pair of mandibles and two pairs of maxillae.
10. Thorax bears 6 pairs of appendages.
11. Behind the appendages is a penis.
12. Abdomen is reduced to a process without any appendages.
13. It is bisexual, paired testes lie at the side of digestive tract and ovaries in the pedicle.
14. Development indirect through nauplius which moults into cypris stage which metamorphoses into adult.

BALANUS

The systematic position is the same as that the *Lepas*.

1. It is found attached to rocks and floating woods, shell of molluses, carapace of turtle etc. in sea water.
2. It has cosmopolitan distribution.
3. It is commonly known as 'rock - barnacle' or 'acorn - barnacle'.
4. It is sessile i.e. without pedicle, therefore, the body is directly attached to the rock or substratum.
5. Body is enclosed in a cylindrical case formed of six calcareous plates - unpaired carina and rostrum and two pairs of carino-laterals.

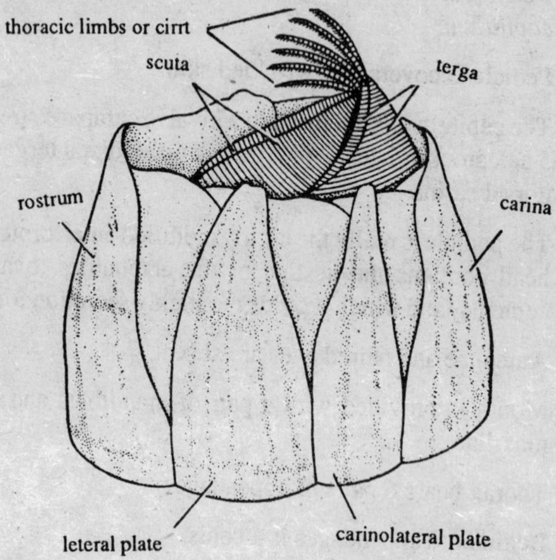

Fig. 1.11. *Balanus*

6. At the top of the shell there is an opening which is closed by two pairs of valves called *opercular valves*.
7. From the opening 6 pairs of thoracic feet which sweep in food particles of animal.
8. Eyes and Entennae are absent.
9. Bisexual or hermaphrodite.
10. Development indirect and a nauplius larva is set free which moults into cypris stage and finally metamorphoses into adult.

SACCULINA

Phylum	-	Arthropoda
Class	-	Crustacea
Subclass	-	Cirripedia
Order	-	Rhizocephala
Genus	-	*Sacculina*

1. *Sacculina* is parasite on carbs found attached between thorax and abdomen.

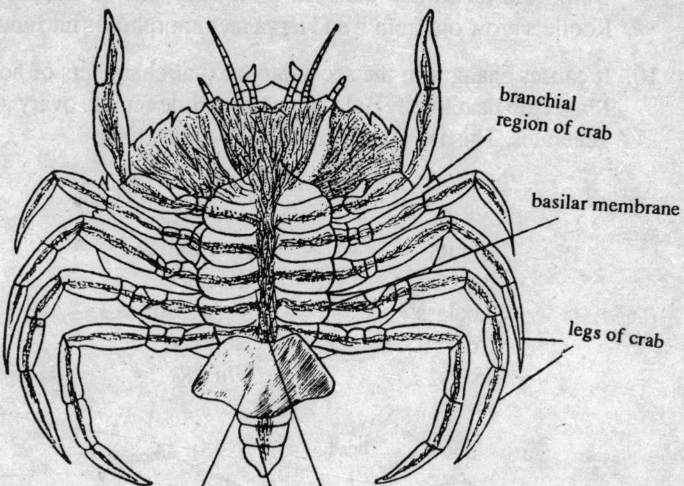

Fig. 1.12 *Sacculina* on the abdomen of crab.

2. It has world wide distribution.
3. It is commonly called 'root-headed barnacle'.
4. Adult is extremely degenerate and appears as a fleshy tumour on the host.
5. Body without carapace, segmentation, appendages and alimentary canal.
6. The peduncle gives out many delicate root-like proccess into the tissue of host through which it absorbs nourishment.
7. Opening of mantle cavity at the posterior end of the body is known as cloacal apertuse.
8. These are bisexual.
9. Development is indirect through cirripide nauplius, which transform into cypris. Cypris attaches itself to the crab by its anten-

nules. The cypris moults and called kentrogen larga. It travels within the blood and attached to ventral side of intestine . Rootlets grow out from it and larva metamorphoses into adult.

10. It causes changes in the secondary sexual characters of host. Male crab develops female characters while female ovary degenerates. This is known as *Parasitic Castration*.

ONISCUS

Phylum	-	Arthropoda
Class	-	Crustacea
Subclass	-	Malacostraca
Order	-	Isopoda
Genus	-	*Oniscus*

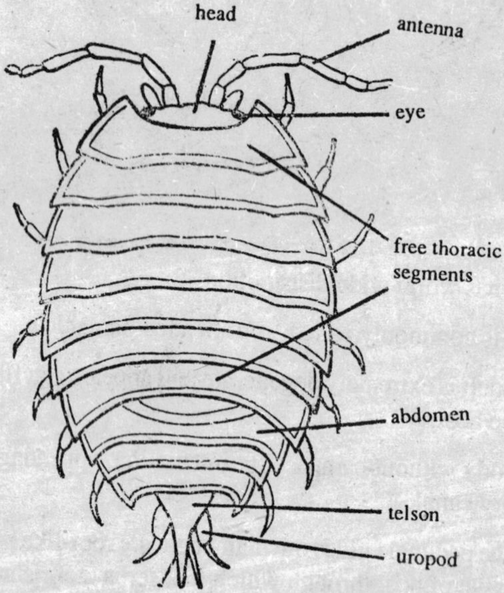

Fig. 1.13 *Oniscus*

1. It is found Inder stones, woods, etc, in damp soil. It is found in Europe.

2. It is commonly called as ‘wood lice’.

3. The body is oval and dorso-ventrally flatenned.

4. First thoracic segment fused with the head to form cephalothorax.
5. Unstalked compound eyes are present.
6. Antennules and antennae are uniramous.
7. The abdominal legs are adapted for aerial respiration and their endopodiates bear delicate branchiae traversed by minute tube called pseudotracheae.
8. Abdomen short.
9. Yongs develop in a brood pouch, present on ventral side.

SQUILLA

Phylum	-	Arthropoda
Class	-	Crustacea
Subclass	-	Malacostraca
Order	-	Stomatopoda
Genus	-	*Squilla.*

1. It is found in burrows of the sand or mud at the bottom of sea.
2. It is found in India, southern coast of Florida and Gulf of Mexico.
3. Body is elongated measuring 25 cm. in length, divisible into head, thorax and abdomen.
4. Carapace is thin and uncalcified. It is produced anteriorly into a movable rostrum.
5. Compound eyes are stalked.
6. Anterior 5 pairs of thoracic appendages are uniramous and directed forwards to act as maxillipedes. The remaining thoracic appendages are biramous and act as legs.
7. The 2nd pair of maxillipedes are very large and form powerful prehensile weapons.
8. Abdomen is relatively large and has a broad telson bearing 3 pairs of marginal spines.
9. Pleopods are biramous and large. They bear gills on their exopodites.

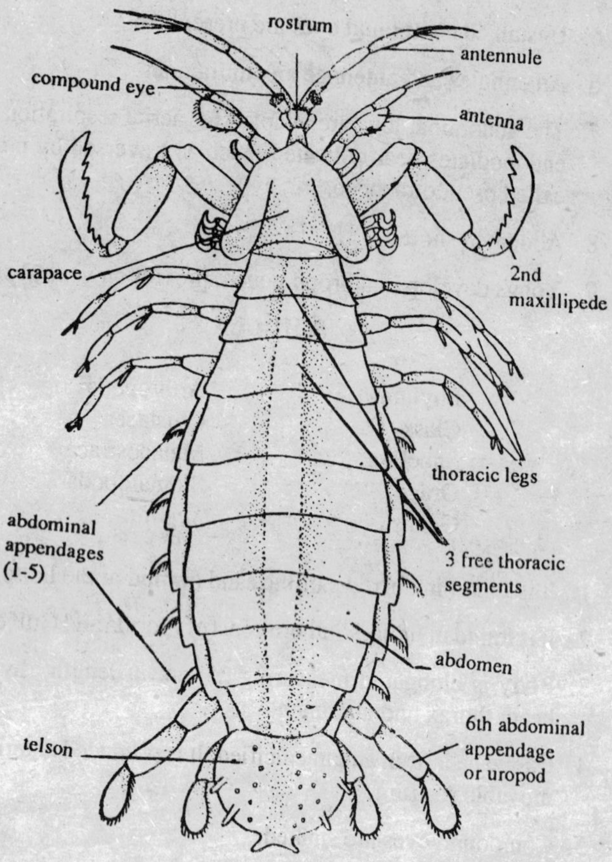

Fig. 1.14 *Squilla*

10. Sixth form large uropods which are flaterred and form the tail-fan along with the telson.
11. The eggs ate carried about by the last three pairs of maxillipedes.
12. The larvae are trnsparent pelagic Alima, which metamorphoses into adult *Squilla*.

ASTACUS

Phylum	-	Arthropoda
Class	-	Crustacea
Subclass	-	Malacostraca
Order	-	Decapoda
Genus	-	*Astacus*

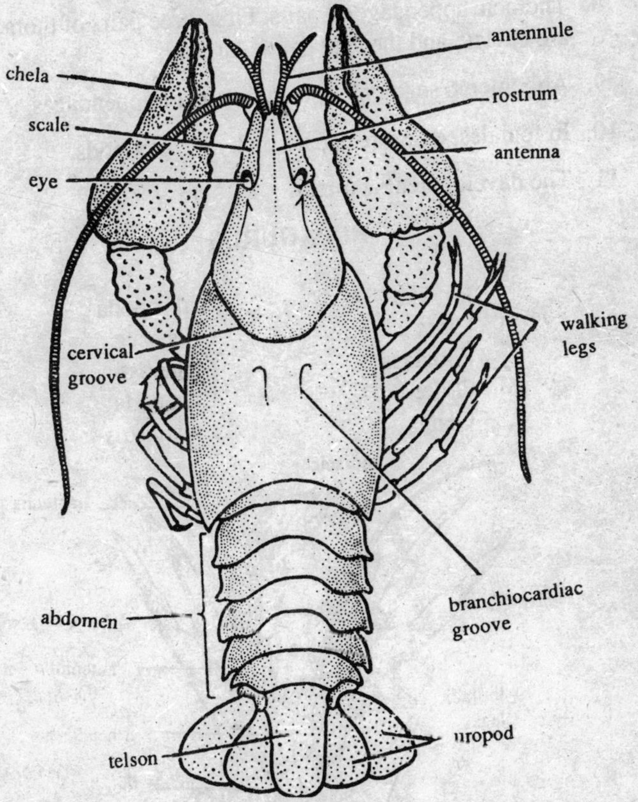

Fig. 1.15 *Astacus*

1. *Astacus* is found in freshwater ponds, streams and lakes all over the world.
2. It is commonly called 'cray-fish'.
3. The body is long and robust, somewhat compressed.

4. Body is divisible into cephelothorax and abdomen.
5. The line of demarcation between the head and thorax is possibly indicated by cervical groove.
6. Rostrum is short and without serrated margins.
7. Head appendages 5 pairs. Eyes, antennules and antennae are present.
8. Thoracic appendages 8 pairs. First three pairs of thoracic legs are chelate and the first pair is largest.
9. Abdomen 6 segmented, with 6 pairs of appendages.
10. In females, eggs are carries on by the pleopods.
11. The development is direct.

EUPAGURUS

Phylum	-	Arthropoda
Class	-	Crustacea
Subclass	-	Malacostraca
Order	-	Decapoda
Genus	-	*Eupagurus*

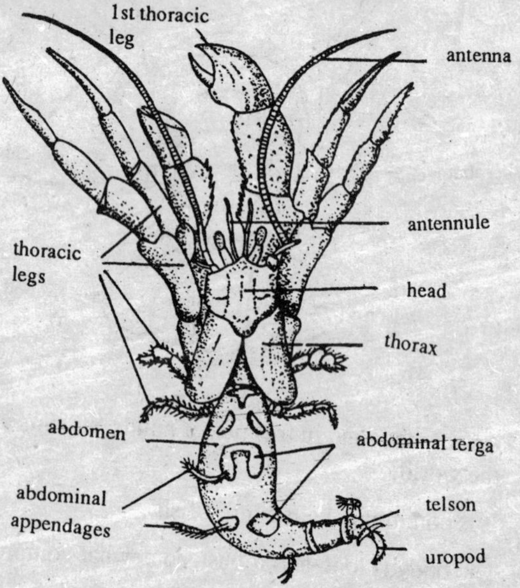

Fig. 1.16 *Eupagurus*.

1. It is found in the empty shells of gastropod molluses.
2. It has cosmopolitan distribution.
3. It is commonly called 'hermit crab'.
4. The body is highly asymmetrical and consists of cephalothorax and abdomen.
5. Cephelothorax is broad and flatenned bearings short antennules, antennae and stalkes eyes.
6. Of the 5 pairs of thoracic legs, first are chelate, 2nd and 3rd are nonchelate, 4th and 5th are chelate but reduced.
7. Abdomen is soft, cylindrical, coiled and unsegmented. It bears reduced appendages on the left- side and none on right side, Uropods are, however, well developed.
8. Swimmerets are found only in females for carrying eggs.
9. They show *commensalism.* The close association between the hermit crabs and sea-anemone in which both are benefetted by each other.

HIPPA

Phylum	-	Arthropoda
Class	-	Crustacea
Subclass	-	Malacostraca
Order	-	Decapoda
Genus	-	*Hippa*

1. It is found in the sand near the sea. It is commonly found on Pacific Coast, U.S.A. and Oregon to Mexixo.
2. It is commonly known as 'mole - crab'
3. Body is ovate with large cephatothotrax.
4. Carapace is smooth.
5. Mouthparts are poorly developed.
6. Head bears long stalked eyes, a pair of biramous antennules and a pair of uniramous antennae.

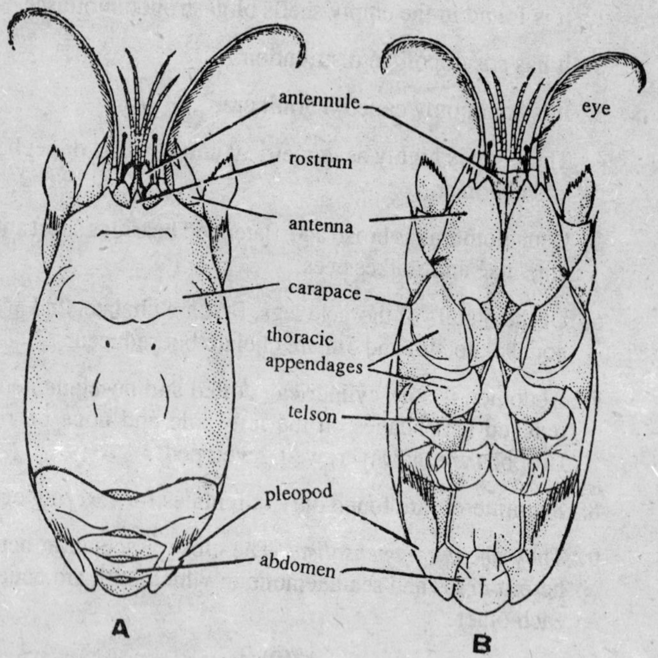

Fig. 1.17. *Hippa*. A = Dorsal view. B = Ventral view.

7. Rostrum is simple and pointed.
8. Thoracic appendages 7 pairs, walking legs 5 pairs adapted for digging. First two pairs of walking legs partially chelate.
9. Abdomen 6 segmented first three pairs of abdominal appendages modifies for swimming and last three pairs are adapted for darting.
10. Feeds like an earthworm, swallowing sand.

CARCINUS

Phylum	-	Arthropoda
Class	-	Crustacea
Subclass	-	Malacostraca
Order	-	Decapoda
Genus	-	*Carcinus*

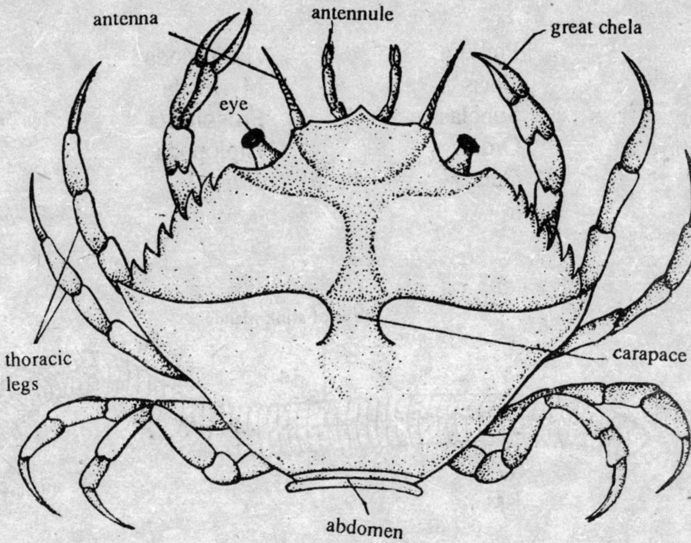

Fig. 1.18 *Carcinus*

1. It is commonly found buried among rocks and in sand near about the water. It has cosmopolitan distribution.
2. Cephelophorax broad and somewhat oval with a broad carapace.
3. The antennae and antennules are short.
4. The eyes are stalkes and lodged in a pit of the cerapace.
5. Third maxillipedes broad, flat and plate - like.
6. Walking legs 5 pairs, the first pair is large chelate.
7. Abdomen reduced and is permanently flexed below cephalothorax.
8. Males with 2 pairs of pleopods, whereas female possesses 4 pairs of them. Uropods absent.
9. Eggs are carries by the females which are attached to the pleopods.
10. Development indirect. Egg develop into zoea which passes into megalopa and finally metamorphoses into adult.

JULUS

Phylum	-	Arthropoda
Class	-	Myripoda
Subclass	-	Progoneata
Order	-	Diplopoda
Genus	-	*Julus*

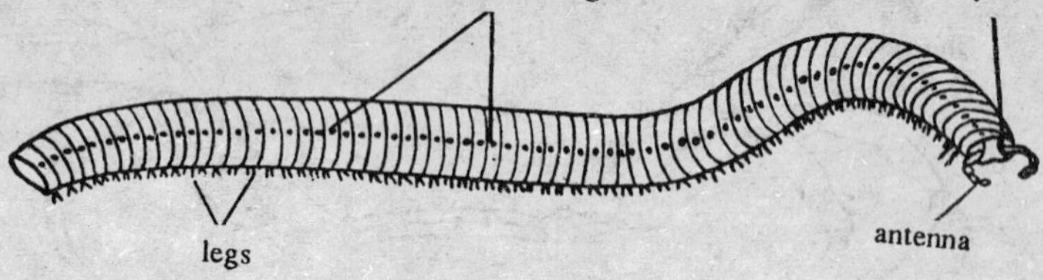

Fig. 1.19. *Julus*

1. It is found is damp and dark places under stones and decying leaves. When disturbed usually roll up into tight coil and emits an unpleasing odour from segmental sent glands.
2. It has cosmopolitan distribution.
3. It is commonly called as 'wire worm'
4. Body cyclindrical and metamerically segmented; differentiated into head and trunk.
5. Head small; antennae short generally 7 segment. One pair of maxillae and mandibles forming *gnathochilarium.*
6. Eyes are minute
7. Trunk cylindrical consisting of very closely ring like segments from 25 to more than 100 segments.
8. In each segment there are two pairs of legs. Since it has numerous legs it is called millipede.
8. Poison glands and poison claws absent.
10. Vascular system well developed.

11. Respiration by tracheae.
12. Sexes are separate.
13. Genital openings situated at the anterior end of body.
14. In male one or both pairs of legs of 7th segment are modified to form copulatory organs.
15. Herbivorous and even scavanger.

SCOLOPENDRA

Phylum	-	Arthropoda
Class	-	Myripoda
Subclass	-	Opisthogoneata
Order	-	Chilopoda
Genus	-	*Scolopendra*

1. It is found under stones, in cervices and other dark places. It is found in India and America.
2. It is commonly called 'Centipede'
3. Body is elongated and dorsoventrally flattened with numerous segments.
4. Head is dintict and bears a pair of antennae, a pair of mandibles and two pairs of maxillae.
5. Trunk segments numerous, each bearing a single pair of legs.
6. First pair of trunk appendages or maxillipedes bears a sharp claw connected with the poison gland.
7. The last pair of legs are usually longer than the rest.
8. The anus lues at the posterior end.
9. Respiration os affected by tracheal tube..
10. The ,vascular system well developed.
11. Sexes are separate.
12. The genital opening situated at the hind end of the body.
13. Bite is dangerous to man and sometime proves to be fatal.

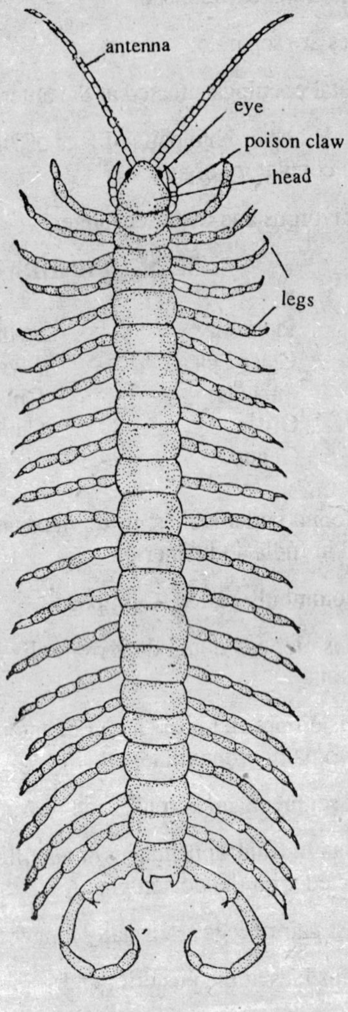

Fig. 1.20 *Scolopendra*--the centipede.

LEPISMA

Phylum	-	Arthropoda
Class	-	Insecta
Subclass	-	Apterygota

Order - Thysanura
Genus - *Lepisma*

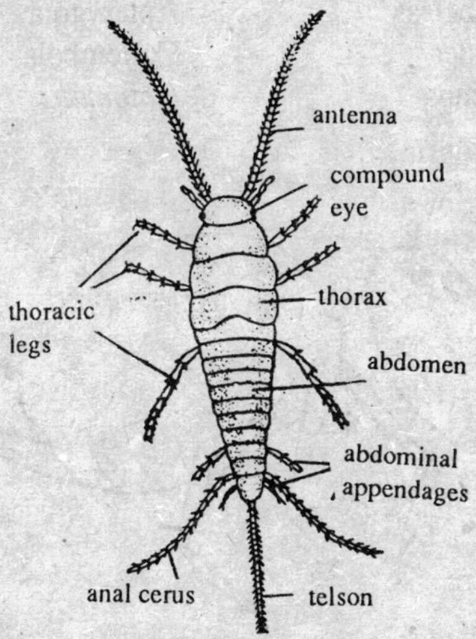

Fig. 1.21 *Lepsima*

1. It is found in houses, in books, papers, clothes, under pictures, in warm countries. It feeds on starch.
2. It is commonly known as 'silver fish'
3. Body is flattered and covered with silvery, scales.
4. Body divisible into head, thorax and abdomen.
5. Head small with a pair of filiform antennae and eyes.
6. Mouth parts chewing type.
7. Thorax three segmented and bears 3 pairs of legs wings are absent.
8. Adbomen 11 segmented with 3 anal cerci and a number of small delicate appendages.
9. Development direct without metamorphosis.

ISOTOMA

Phylum	-	Arthropoda
Class	-	Insectsa
Subclass	-	Apterygota
Order	-	Collembola
Genus	-	*Isotoma.*

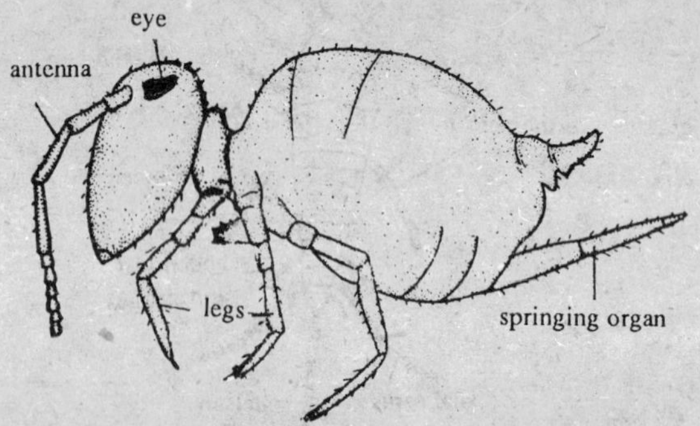

Fig. 1.22 *Isotoma*

1. There are found in damp places, under leaves, cerives of bark.
2. It is commonly called 'Spring tail'.
3. Body is divisible into head, thorax and abdomen.
4. Head with 4 jointed antennae.
5. Compound eyes are absent, O celli scattered.
6. Thorax 3 segmented with 3 pairs of legs. Wings absent.
7. Abdomen 6 segmented. First segment with a tubular adhesive organs, *collophore*. 3rd segment with *tentaculum* and 4th with *furcula* (spring organ).
8. It secretes glue-like substance used for sticking body to the object.

SCHISTOCERCA

Phylum	-	Arthropoda
Class	-	Insecta

MANTIS

Phylum	-	Arthropoda
Class	-	Insecta
Subclass	-	Pterygota
Division	-	Exopterygota.
Order	-	Mantodea
Genus	-	*Mantis*

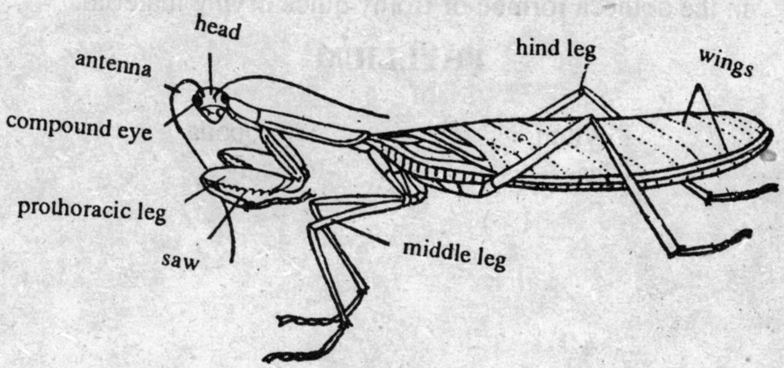

Fig. 1.26 Mantis

1. These are arboreal feeding on living prey (carnivorous).
2. It is found in North America, U.S.A. Africa, South Europe, Western Asia.
3. It is commonloy known as 'preying mantis'
4. Body is long, slender, usually green in colour.
5. Body is divisible into head, thorax and abdomen.
6. Head is small, triangular with vertical face and freely movable on slender neck.
7. Large compound eyes, three ocelli and long filiform antennae are present on head.
8. Mouthparts are chewing type.
9. Prothorax is much elongated and prothoracic legs are modifies

for grasping and holding thew prey.

10. Wings are well developed. These are folded flat and overlapping the sides of the body.
11. Abdomen 10 segmented. Ovipositor is no exected.
12. Nineth segment in makes bear a pair of styli.
13. Short flattered and segmental ceric present.
14. Female eats the male after muptial. 200 or more eggs are enclosed in the ootheca formed of frothy quick drying material.

PHYLLIUM

Phylum	-	Arthropoda
Class	-	Insecta

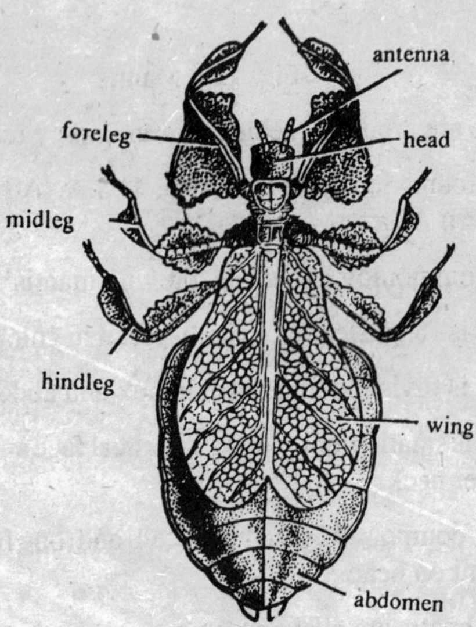

Fig. 1.27 Phyllium

Subclass	-	Pterygota
Division	-	Exopterygota
Order	-	Phasmida
Genus	-	*Phyllium*

1. It is found in the moist part of Oriental region, Forests of Assam etc.
2. It is commonly called 'leaf - insect'
3. Body is broad and flatenned.
4. Body is divisible into head, thorax and abdomen.
5. Head is small bears small antennae.
6. Thorax is also small and bears 3-pairs of green coloured legs.
7. Tegmen or first pairs of wings are small than the other pair, having leaf-like colouration and veination.
8. Sexes distinct. Male with small tegmina, well developed wings and narrow bodies. Female with extensive tegmina produced into leaf like expansions.
9. It is an example of *mimicry* or *Camouflage* in which protective resemblance in both form and colouration, is similar to the objects in their surroundings.

CARAUSIUS

Phylum	-	Arthropoda
Class	-	Insecta
Subclass	-	Pterygota
Division	-	Exopterygota
Order	-	Phasmida
Genus	-	*Carausius*

1. It is found in tropical forests and is asboreal , it is found in India, Sri Lanka, Burma, Asia, Europe.
2. It is commonly known as 'stick--insect'.
3. Body is elongated and slender with long legs and antennae resembling a twing or stick or any other foliage on which they feed.

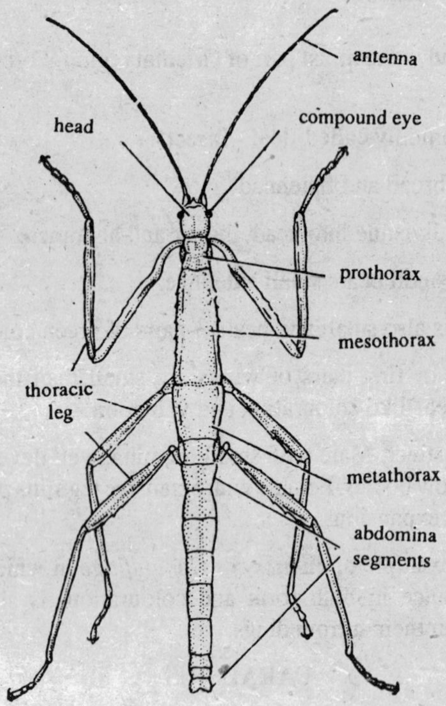

Fig. 1.28 *Carausisus*

4. Head is small with a pair of filiform antennae and compound eyes. Mounthparts chewing type.

5. Thorax is elongated bearing 3 pairs of legs and the wiongs are absent.

6. Abdomen 10 segmented.

7. Cerci small and unsegmented.

8. Sexes are separate, Male is small, active and winged. Female is large, sluggish and apterous.

9. Exclusively parthenogenesis.
10. Power of regeneration or autotomy is very great.

FORFICULA

Phylum	-	Arthropoda
Class	-	Insecta
Subclass	-	Pteygota
Division	-	Exopterygota
Order	-	Dermaptera
Genus	-	*Forficula*

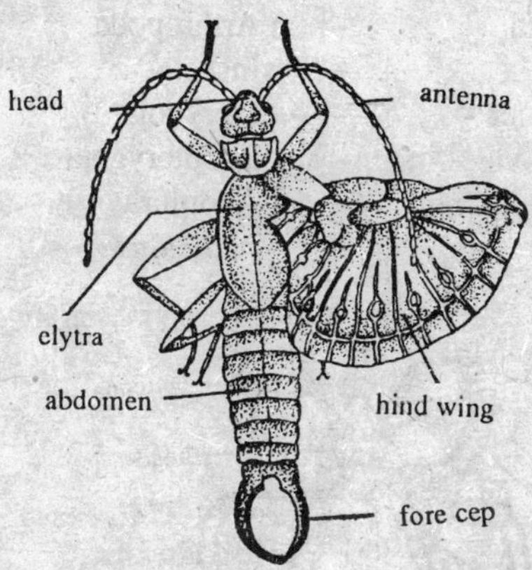

Fig. 1.29 *Forficula*

1. Found under stones, moist places under stones in decaying vegetable matter etc.
2. It is found in Europe, N. America, Mexico, Asia, Africa and on the coasts of Indian ocean.
3. It is commonly known as 'earwing'.
4. Small insect with elongate body.
5. Body is divisible into head, thorax abd abdomen.
6. Head contains a pair of eyes and a pair of long--filiform antennae.
7. Mouth parts chewing type.

8. Thorax contains 3 pairs of walking legs and peculiar hind-wings.
9. The forewings form hard veinless truncated wing cover below which the hindwings are present.
10. Anal cerci are modified into a pair of *forceps* which is raised upwards and is opened widely in a threatening manner, when the animal is disturbed.
11. Ovipositor absent.
12. Metamorphosis gradual and it exhabits parental care for eggs and young ones.

PEDICULUS

Phylum	-	Arthropoda
Class	-	Insecta
Suborder	-	Pterygota
Division	-	Exopterygota
Order	-	Anoplura
Genus	-	*Pediculus.*

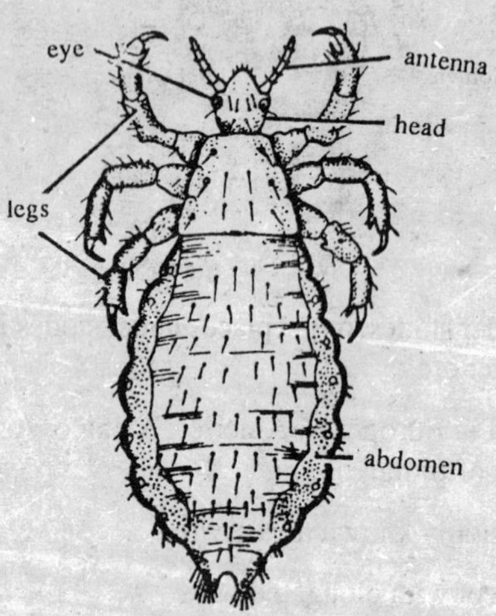

Fig. 1.30 *Pediculus*

1. These are minute, flat, wingless and actoparasite of man. It has cosmopolitan distribution.

2. It is commonly known as 'human louse'.
3. Body is dorso-ventrally flattered and pale in colour with dark marking along the sides.
4. Head bears a pairs of antennae which is 5 segmented, and a pair of compound eyes.
5. Mouthparts piering and sucking type.
6. Thoracic segments are fused in one mass and contain 3 pairs of clawed legs, which form clinging organs with the tissue of the host.
7. Abdomen is 9 segmented.
8. Males are smaller than females and their posterior ends turned upward.
9. Metamorphosis is absent.
10. It is a major factor in transmitting relapsing fever, typhus and trench fever. It sucks the blood of man.

EPHEMERA

Phylum - Arthropoda

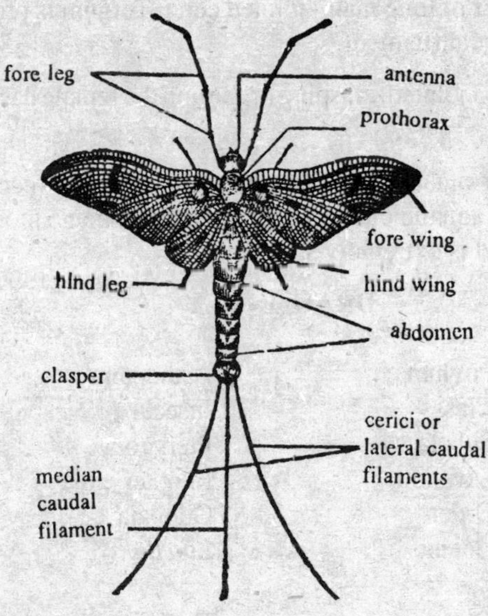

Fig. 1.31 *Ephemera*

Class	-	Insecta
Subclass	-	Pterygota
Division	-	Exopterygota
Order	-	Ephemerida
Genus	-	*Ephemera*

1. It is a soft-bodies and most familiar insect found on the margins of streams, lacks etc.
2. It is found in U.S.A., India, Sri Lanks and Burma.
3. It is commonly called as 'Mayfly'.
4. Body is divisible into head thorax and abdomen
5. Head is small contain small setose antennae, compound eyes and 3 ocelli.
6. Mouthparts vestigial.
7. 3 pairs of inconstant walking legs.
8. Two pairs of membranous wings, forewings larger than hind-wings.
9. Abdomen is 10 segment with a reduced tergum. Last segment bears a pair of long many jointed cerci. Tergum is produced into a long caudal filament.
10. Males has 3 jointed, clasping organ and the female has no oripositor.
11. Nymph or naiad lives for a long time upto 3 years in water, feeding in aquatic plants and animals. The imago moults prior to mating and is extremely short lived.

DRAGON FLY

Phylum	-	Arthoptera
Class	-	Insecta
Subclass	-	Pterygota
Division	-	Exopterygota
Order	-	Odonata
Genus	-	*Libellula*

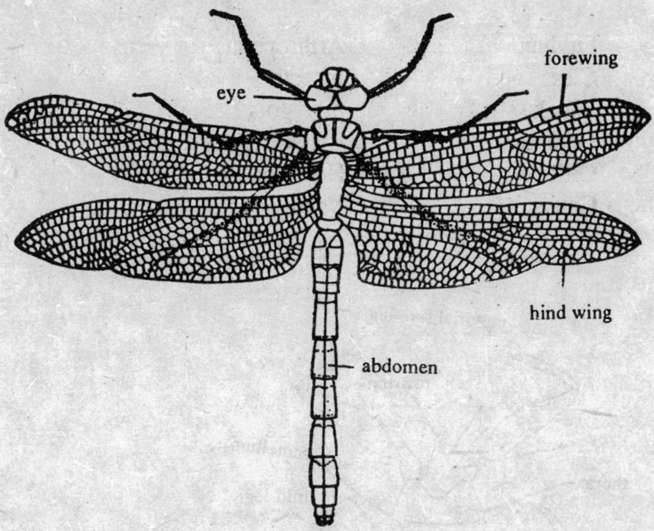

Fig. 1.32 *Libellula*

1. It is of world wide distribution and generally found darting balancing in the air.
2. Body is slender and divisible into head, thorax and abdomen.
3. Head bears a pair of large compound eyes and small inconspicuous antenne.
4. Mouth parts chewing type.
5. Prothorax is small. Meso- and metathorax are fused. Legs are weak and used for resting.
6. Both pairs of wings are well developed and membranous. These are held in a horizontal position.
7. The abdomen in elongated often slender made up of 10 segments. The genital openings in both, male and female, lies at the hind end. Male copulatory organs on the 2nd and 3rd sternite.
8. Nymphs are aquatic, breath by rectal gills and feeding upon quatic insects and organisms.

NEPA

Phylum	-	Arthropoda
Class	-	Insecta
Subclass	-	pterygota
Division	-	Exopterygota
Order	-	Hemiptera
Genus	-	*Nepa*

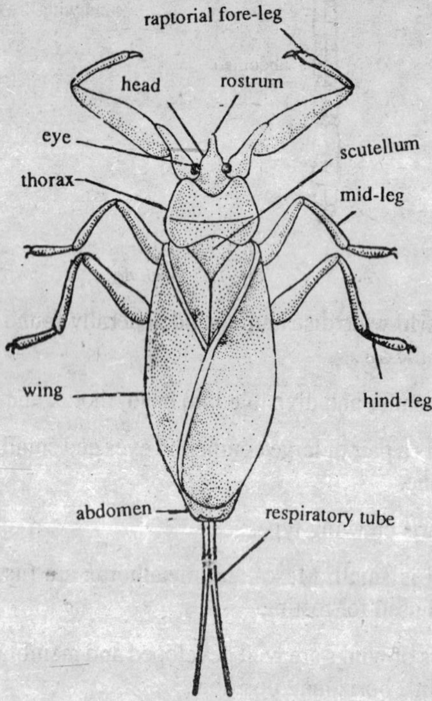

Fig. 1.33 *Nepa*

1. It is commonly found in shallow water. It has world wide distribution.
2. It is commonly called ' water scorption'

3. The body is dorsoventrally flattenned and divisible into head, thorax and abdomen.
4. Head comprises a pair of 3 jointed antennae and compound eyes.
5. Mouth parts form a long tostrum adapted for piercing and sucking.
6. The respiratory tube is most peculiar. It consists of two spine-like process.
7. There are three pairs of false spiracles situated on 3rd, 4th and 5th ventral abdominal segments respectively.
8. The false spiracles comprise sieve-like structures with the perforations occuleded by a delicate membrane.
9. Sexes are separate.
10. The female is provided with a pointed toothed ovipositor.
11. The eggs are deposited in chains. The ova adhere to one another by means of sevan long filaments radiating from on extremity.

BELOSTOMA

The systematic position is the same as that of *Nepa*

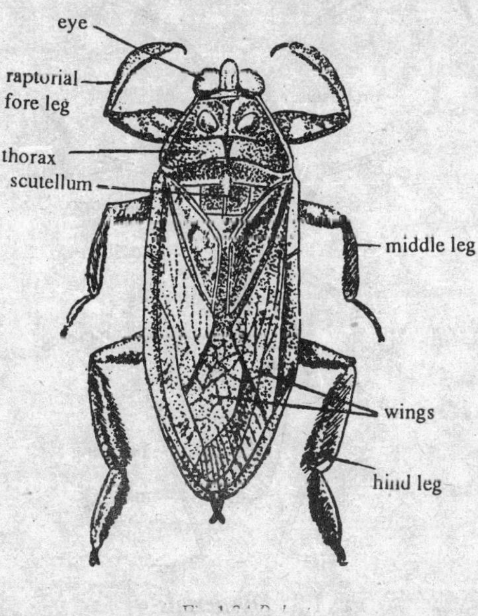

1. It is the largest insect found in rivers, ponds, lake, feeding on insects, tadpoles, snails etc.
2. It is found in India, U.K., North America and South Africa.
3. It is commonly called 'gaint water bug'
4. Body is dorsoventrally flatenned and divisible into three parts, head, thorax and abdomen.
5. Head is small with a pair of large eyes and 4 jointed antennae. Mouth parts piercing and sucking type.
6. The Thorax bears 3 pairs of legs. The mid legs and hindlegs are flatenned and adapted for swimming the forelegs raptorial.
7. First pair of wings hemielytra and second pair membranous.
8. The abdomen is provided with two retractile apical appendages.
9. Ovipositor well developed. Females of some species deposit their eggs on the back of male (parental care).

DYSDERCUS

The systematic position is the same as that of *Nepa*

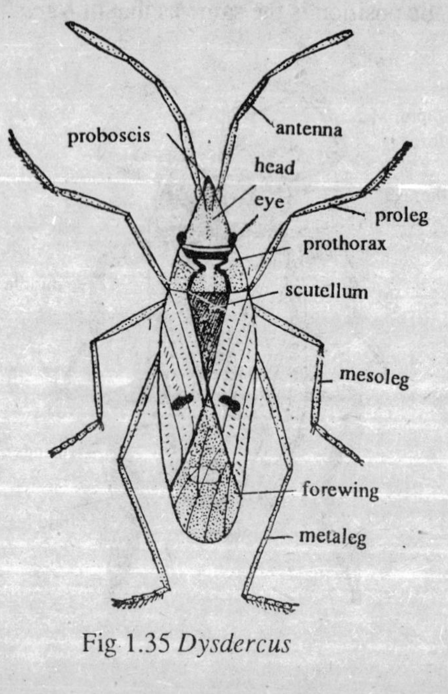

Fig 1.35 *Dysdercus*

1. It is a common pest of cotton and commonly known as 'Red cotton-bug'.
2. Body is black and red in colour.
3. Body divisible into head thorax and abdomen.
4. Head bears a pair of compound eyes and 4 jointed antennae.
5. Mouth parts piercing and sucking type.
6. Wings two pairs, basal half of the forewings hard and leathery (hemelytra), Hindwings soft and membranous.
7. Triangular scutellum present between the leases of wings.
8. Pronotum large.

CICADA

The systematic position is same as that of *Nepa*.

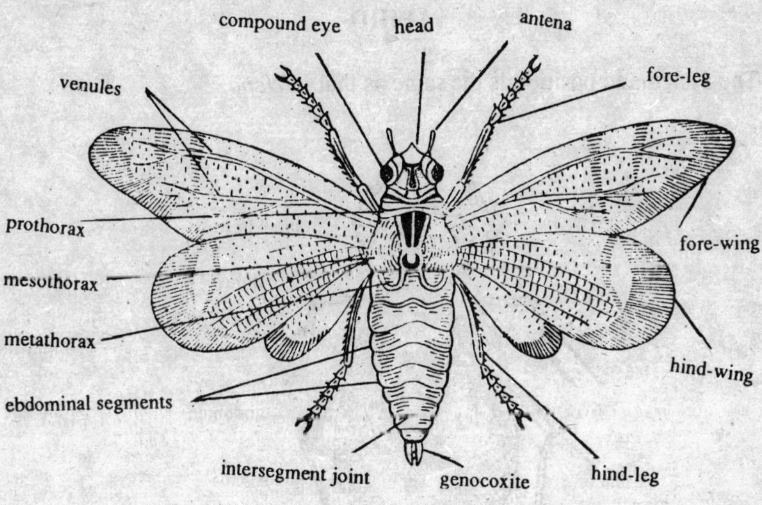

Fig 1.36 *Cicada*

1. It is found in the forests and moist places, abundant in tropical, subtropical and temperate regions.
2. Body is divisible into head, thorax and abdomen.

3. The head is broad possessing two lateral compound eyes and two very short antennae.
4. The thorax comprises prothorax, mesothorax and metathorax.
5. There are two pairs of wings: fore-wings are larger and somewhat triangualr and the hind-wings are oval.
6. The andomen is elongated and consists of ten segments.
7. The sound production habit of this insect is very peculiar.
8. The sound is produced only by the male which possesses a drum-like apparatus called aedeaugs.
9. Metamorphosis imcomplete, egg hatches into nymph.
10. The life-history is very interesting as it takes about 13 to 17 years to complete.
11. Adults soon dies after mating.

APHID

The systematic position is the same as that of *Nepa*.

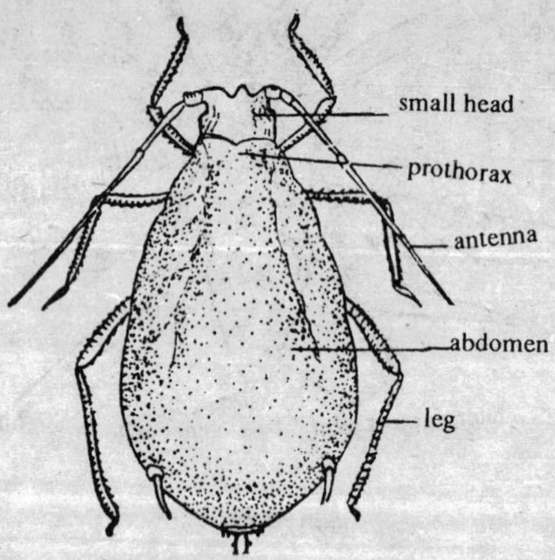

Fig 1.37 *Apple-grain Aphid*

1. It is ectoparasite on plants sucking their juice. It is found specially on the plants of mustard, wheat and cotton.

3. It is commonly called 'plant-lice'.
4. Body is short divisible into head, thorax and abdomen.
5. Head is distinct with long straight antennae. Eyes are small.
6. Mouth parts are piecrcing and sucking type.
7. Thorax and abdomen robust. Abdomen is elongated.
8. Exhibit alternation of sexual and pasthenogenetic phases, females always develop from parthenogenesis.
9. Both apterous and winged forms.

DANAUS

Phylum	-	Arthropoda
Class	-	Insecta
Subclass	-	Pterygota.
Division	-	Endopterygota
Order	-	Lepidoptera
Genus	-	*Danaus.*

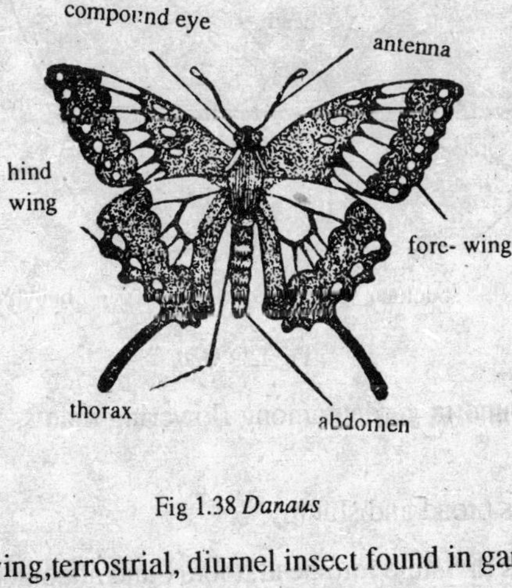

Fig 1.38 *Danaus*

1. It is a flying,terrostrial, diurnel insect found in gardens.
2. Commonly called butter flies.

3. Beautifully and brightly coloured insects.
4. Body is divisible into head, thorax asnd abdomen.
5. Head with a pair of compound eyes and club-shaped antennae.
6. Mouth parts siphoning type. Maxillae are modified to form sucking tube.
7. Prothorax is fused with mesothorax.
8. Two pairs of wings which are typically covered with scales.
9. Abdomen is 10 segmented and covered with scales.
10. Cerci and oripositor are lacking.
11. Development complete.

MOTH

The systematic position is the same as that of *Danaus*.

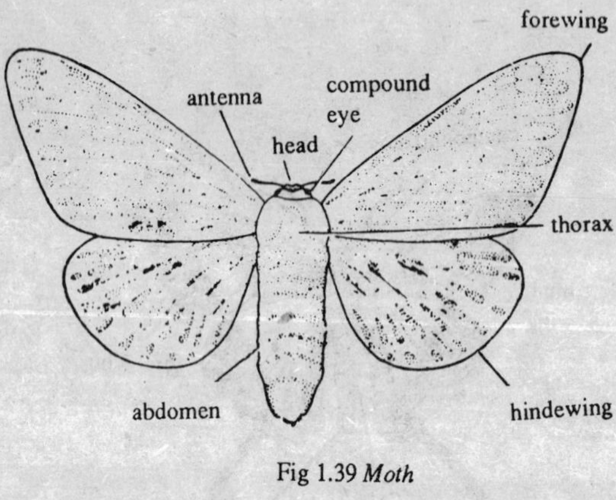

Fig 1.39 *Moth*

1. It is found in garden among flowering plants. It is nocternal in habit.
2. Body is broad and stumpy.
3. Body is divisible into head, thorax and abdomen.
4. Head bears a pair of eyes and antennae. The antennae are usually

feather-like and terminal knobs are absent.

5. Wings two pairs and are held horizontally in resting condition and covered with scales.
6. Mouth parts are siphoning type.
7. Abdomen is 10 segmented.

COCCINELLA

Phylum	-	Arthropoda
Class	-	Insecta
Subclass	-	Pterygota
Division	-	Endopterygota
Order	-	Coleoptera
Genus	-	*Coccinella*

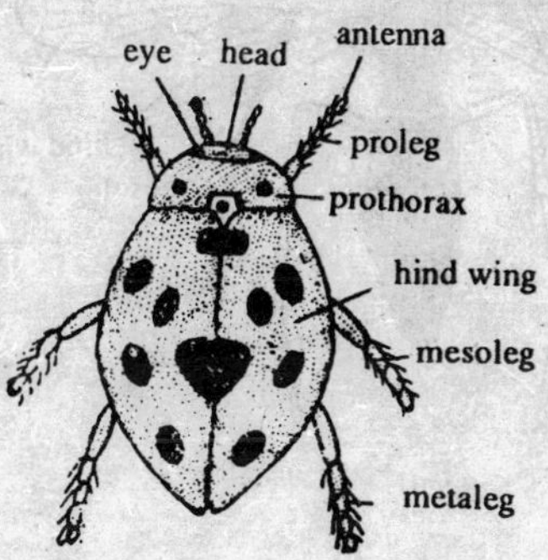

Fib 1.40 *Coccinella*

1. These are commonly called, 'Lady Bird - Beetle'. These are terrestrial animal.
2. Body divisible into head, thorax and abdomen.
3. Head incompletely hidden by pronotum.
4. Head bears compound eyes and a pair of antennae. Mouth parts are chewing type.
5. Large number of species are carnivorous and predacious. Feeds

on aphids, coccids and other soft bodied insects.

VESPA

Phylum	-	Arthropoda
Class	-	Insecta
Subclass	-	Pterygota.
Division	-	Endoptezygota
Order	-	Hymenoptera
Genus	-	*Vespa*

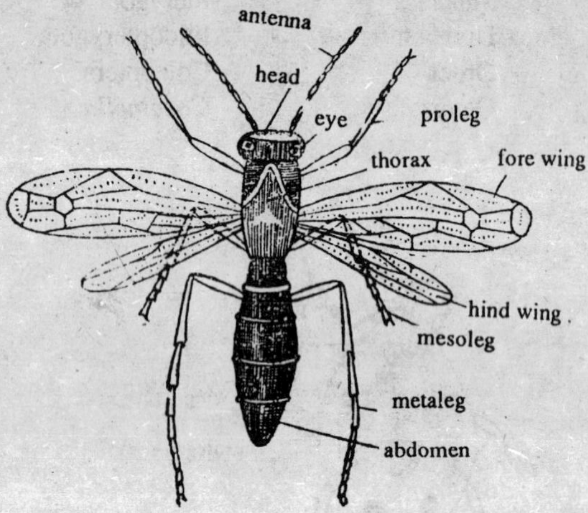

Fig. 1.41 *Vespa*

1. It is an Indian hornet, which is usually chestnut red and makes large nests of mud over walls and trees.
2. Body strongly built, usually chestnut red in colour with yellow, transverse bars.
3. Body is divisible into head, thorax and abdomen.
4. Head is globular concave posteriosly with small ocelli and compound eyes.
5. Thorax deep and cubical wings are membranous. Hind wings short

and bear hooks on their anterior margin which fit into a groove on the posterior margin of forewings.

6. The abdomen petiolate, large with pointed apex.
7. Modified ovipositor as sting is present. It stings painfully and causes local pain and swelling.
8. *Vespa orientalis* is a social colonial insect. Colony consists of queen males and workers. The impregnated female hibernates in winter and forma a new colony at the next spring.

CAMPONOTUS

The systematic position is the same as that of *Vespa*.

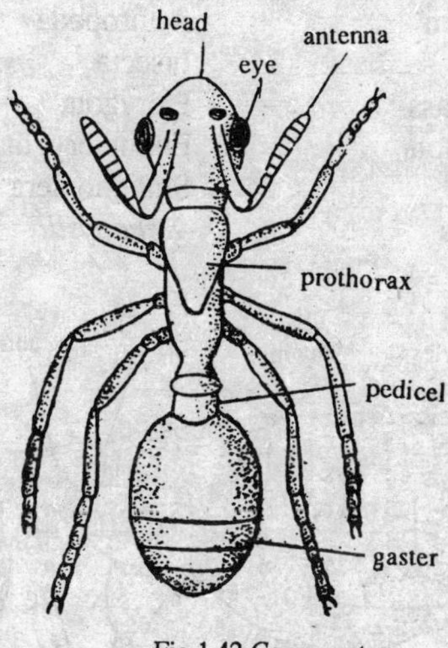

Fig 1.42 *Camponotus*

1. Usually social insects live in burrows, holes, cervices etc.
2. Body divisible into head, thorax and abdomen.
3. Head varies in shape and mandibles are variable in forms.
4. Antennae consits of 4-13 segments. Male has one more segment than the female.
5. Compound eyes and ocelli are well developed in males where as reduced or vestigial in females and workers.

6. Stridulatory (sound producing) organ is present is abdomen.
7. Large and well developed stings are present in females and workers.
8. Mating takes place in winged stage during nuptial flight wings are then lost.
9. These are scavanger but are also harmful to man by destroying grain and seeds.
10. Other commen Indian ants are : *Oceophylla smaragetina* (Weaver), *Holeomyrmex* (harvester) and *Formica* (red ant).

XENOPSYLLA

Phylum	-	Arthropoda
Class	-	Insecta
Subclass	-	Pterygota
Division	-	Endopterygota
Order	-	Siphonoptera
Genus	-	*Xenopysllа*

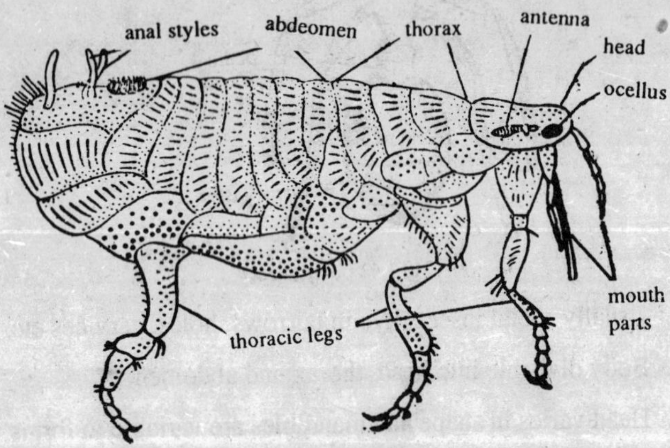

Fig 1.43 *Xenopsylla*

1. It is a sanguivorous ectoparasite of rats, birds and man.
2. It is commonly found in India, Sri Lanka and Malaysia.

3. It is commonly called 'rat--flea'
4. Body is laterally compressed without wings.
5. Integument is covered with spines and bristles.
6. Head broad jointed with small thorax. Abdomen is very large.
7. Antennae are clavate, short and concealed in grooves.
8. Compend eyes and ocelli are absent.
9. Mouthparts piercing and sucking type.
10. Walking legs are elongated, clawed and modified for jumping.
11. It transmits 'plague' from rat to man.

ARANEA

Phylum	-	Arthropoda
Class	-	Arachnida
Order	-	Araneida
Genus	-	*Aranea*

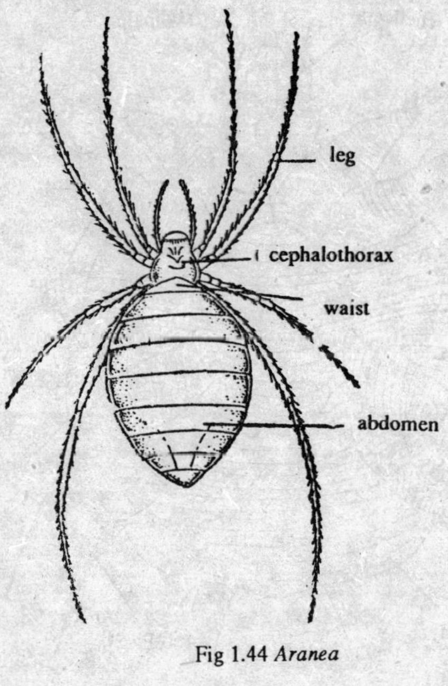

Fig 1.44 *Aranea*

1. It is found in houses and gardens.

2. It is commonly known as 'orb-webbed spider'.
3. Body consists of cephalothorax or prosoma and abdomen or opisthosome, the two being connected by a narrow pedical.
4. Prosome with simple 8 eyes and 6 pairs of appendages. A pair of subchelate *chelicerae* , a pair of nonchelaste *pedipalps* and 4 pairs of walking legs.
5. Opisthosome is unsegmented , bears three pairs of spinnerets which produce thread for making web.
6. Poison glands open on the fangs of chelicerae.
7. Respiration by book lungs and tracheae.
8. Excretion by coxal glands and Malpighian tubules.
9. Sexes are separate. Female is larger than male and eats up after copulation.

SARCOPTES

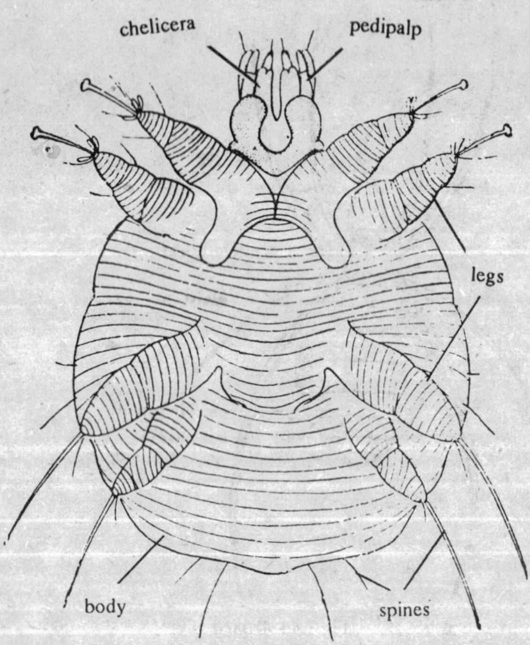

Fig 1.45 *Sarcoptes*

Phylum	-	Arthropoda
Class	-	Arachnida
Order	-	Acarina
Genus	-	*Sarcoptes*

1. It is an ectoparasite on human skin and causes horizontal burrows in region between fingers, elbow etc. Leads to echzema. Commonly known as 'itch--mite of human'
2. It is rounded or oval and dorsoventrally flatenned with transverse striations and short spines.
3. Prosoma and opisthosoma not differentiated.
4. Four pairs of legs. Anterior two legs are stronger and project beyond the body and provided with terminal stalked suckers. Posterior two legs are shorter and attached more ventrally and carry long bristles.
5. Mouth parts borne on capitulum, chelicerae chelate.

IXODES

The systematic position is the same as that of *Sarcoptes.*

1. It is an ectoparasite of sheep and commonly known as 'tick'.
2. Body ovoid, dorsoventrally flatenned and unsegmented.
3. Capitulum projects from the anterior end of the body and bears chelicerae and pedipalpi.
4. Dorsal surface of body is covered by shield-like scutum.
5. Eyes lateral and borne on scutum.
6. Walking legs 4 pairs with adhesive pads and claws.
7. Anus with adanal shield.
8. Female is larger than male.
9. They produce *ixodin* which acts as anesthetic, so that the host does not feel paints.

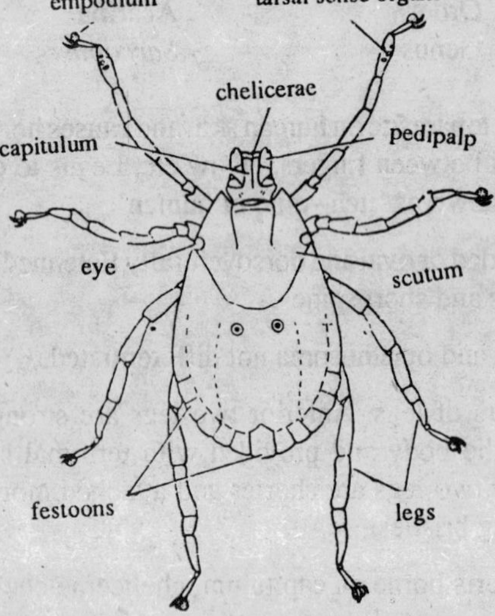

Fig 1.46 *Ixodes*

CHELIFER

Phylum	-	Arthropoda
Class	-	Arachnida
Order	-	Pseudoscorpionida.
Genus	-	*Chelifer*.

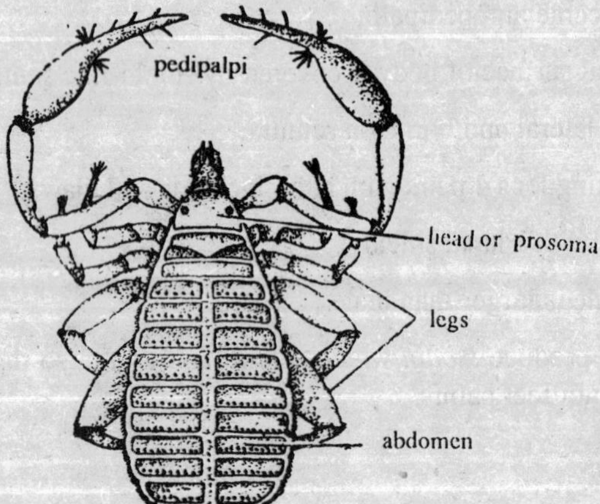

Fig 1.47 *Chelifer*

1. It is found commonly under the bark of trees.
2. Minute animals which have the appearence of scorpion but without tail hence commonly called 'pseudoscorpion'.
3. Body consists of prosoma and an opisthosoma.
4. Prosoma is broad and continuous bearing 6 pairs of appendages.
5. The chelicerae small 2 jointed and chelte.
6. Pedipalpi 5 jointed large and claw-like.
7. A pair of spining glands open on the tips of chelicerae.
8. Abdomen is 12 segmented.
9. Respiration by tracheae.
10. Males have a pair of reversible organ ' ram's horn's, supposed to be copulatory in function.

2

PALAEMON

Palaemon, commonly known as *prawn*, is found in fresh water streams, ponds, rivers etc. It serve as a good example to exhibit the characteristics of class Crustacea.

SYSTEMATIC POSITION

Phylum	-	Arthropoda
Class	-	Crustacea
Subclass	-	Malacostraca
Order	-	Decapoda
Suborder	-	Macrura
Genus	-	*Palaemon*

Habits and habitat.

A number of species of *Palaemon*, such as *P. carcinus, P. malcolmsonii* etc., are found in the fresh water streams, ponds etc. of India, Bangla Desh and distant countries such as West Indies, Mexico and Australia. These animals generally like slow-running waters and are bottom-feeders. They are nocturnal in habit and come out in the night in search of food. The food is generally algae or other water plants and sometimes, insects. *P. malcolmsonii* breeds in May, June and July and the females carry eggs attached to their abdominal appendages.

External Morphology

Shape and size. Prawn is a bilaterally symmetrical animal. The body is elongated and spindle-shaped and is well-adapted for swimming in water. The size of the adult *Palaemon* varies considerably in different species. *P. carcinus* measures about 75 cm in length (found in Travancore). *P. malcolmsonii* from Central India, Tamilnadu and Bengal is about 25-30 cm in length.*P.lamarri*, common in lakes and ponds is quite small about 2.5 -5 cm in length. *P.vulgaris*, a common prawn of America, measurers 30mm in length.

Colouration. The embryonic stages of prawn are usually translucent and whitish in colour. Adult animals of different species are of different colours. It is pale-yellow or greenish with a brown tinge or with orange -red patches or dull pale-blue.

Division of body. The body is divisible into two district regions:-

I. Anterior, immovable and unjointed cephabothorax.

II. Posterior, morable and jointed abdomen.

I. ***Cephalothorax***. It is rigid and immovable. It is straight and oval in cross section. It is composed of 13 segments. In the embryonic condition, cephalothorax consists of first 14 fused segments and a pre-segmental region. Later on during development, first segment of the embryo disappears in the adult, while 13 segments remain. There are 13 pairs of appendages in these thirteen segments. The presegmental region is also found in adults and it possesses a pair of stalked compound eyes.

The cephalothorax is covered by a large chitinous shield, comprises two parts: *dorsal shield* and the *sternal plates*.The dorsal shield further consists of anterior *dorsal plate* and posterior *carapace* but there is no line of demarcation between. The dorsal plate is triangular and it extends forwards as a laterally compressed process, the *rostrum*. The upper and lower margins of the rostrum are serrated and densly fringed with setae. At the base of rostrum, on either side, there is an *orbital notch* to which movable jointed stalk of the compound eyes is attached. Just below each orbital notch, the dorsal plate bears a prominant *antennal spine*. Behind each antennal spine there is a relatively small *hepatic spine*.

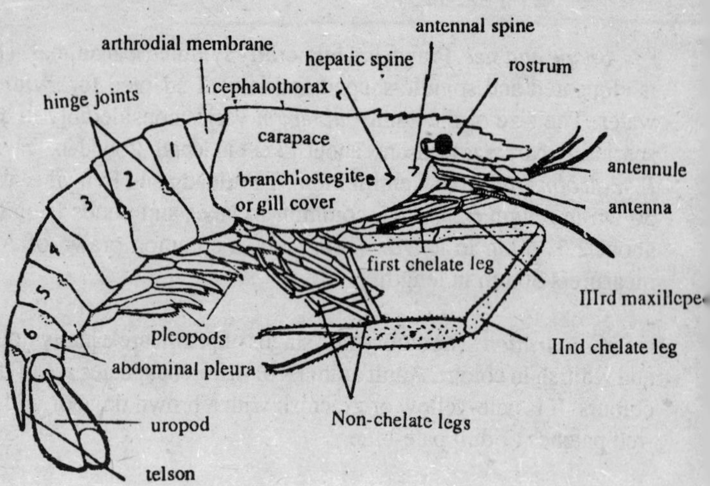

Fig. 2.1 *Palaemon* Lateral view

The carapace covers the thorax dorsally and laterally and represents the fused thoracic terga and pleura. On either sides of the body, the carapace hangs freely as a convex gill cover or *branchiostergite*. These enclose a narrow gill-chamber, which protects gills beneath them.

II.***Abdomen***. It consists of 6 posterior movable segments and a terminal conical structure, the *telson*. The telson articulates with the 6th abdominal segment by a pair of hinge - joints.

Each abdominal segment is covered dorsally by a convex sclerite, the *tergum* and ventrally by the *sternum*. The tergum extends downwards on either side as a free flap, the *pleuron*. The pleuron covers the base of the appendage of its side, and is connected with it by a small sclerite, the *epimeron*. Pleura of the 2nd abdominal segment are largest and differ from others in overlapping the pleura of both the 1st and the 3rd segments. Pleura of the 6th segment are small and point backwards. The adjacent terga and sterna are connected by flexible *arthroidal membrances* of uncalcified cuticle. The adjacent terga articulate by a hinge-joint on either side. However, the hinge joint between the 3rd and 4th segments are lacking. These joints allow only up and down movement of the segments.

External apertures- Mouth is silt-like aperture opening mid-ventrally at the anterior end of the cephelothorax. It is surronnded by maxillipedes and maxillae. *Anus* is in the form of a longitudinal aperture situated ventrally at the base of telson. One pair of *renal apertures* open on raised papillae on the inner surface of the antennae. Sexes are separate. Paired *female genital apertures* are located on the inner surface of coxae of the third pairs of walking legs in female. The paired *male apertures* are located on the inner surface of coxae of the 5th pair of walking legs in male. The openings of the statocysts open in a deep depression dorsally on the basal segment of each antennule.

Appendages

The appendages of prawn show considerable diversity in structure. In different regions of the body appendages are well-adopted for different functions. These appendages are homologous as all have the same typical structure and development.

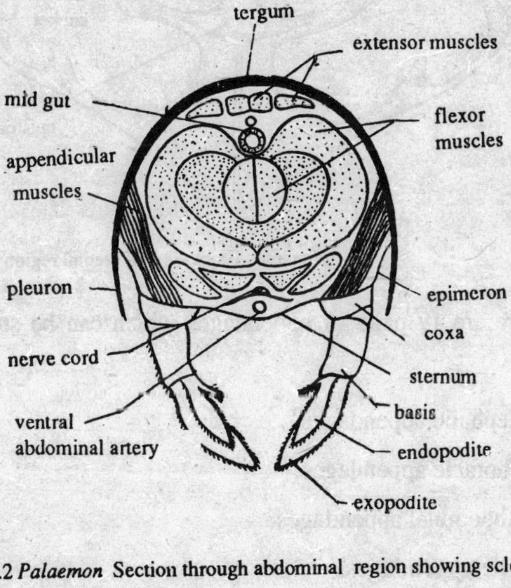

Fig. 2.2 *Palaemon* Section through abdominal region showing sclerites

Appendages appear very different from one another can be reduced to a general type of appendage called, *stenopodium*. A stenopodium consists of two parts (i) *protopodite* or *sympod or stalk* and (ii) *rami* supported on the protopodite. The protopodite is a 2 jointed structure the proximal podomere with which the appendage is attached to the body is *coxopodite or coxa* and the distal joint is known as *basipodite or basis.* Rami are leaf like structures the outer one is called *exopodite* and inner one is *endopodite*. When there is only one ramus the appendage is called *uniramous* and when both are present it is a *biranous* appendage.

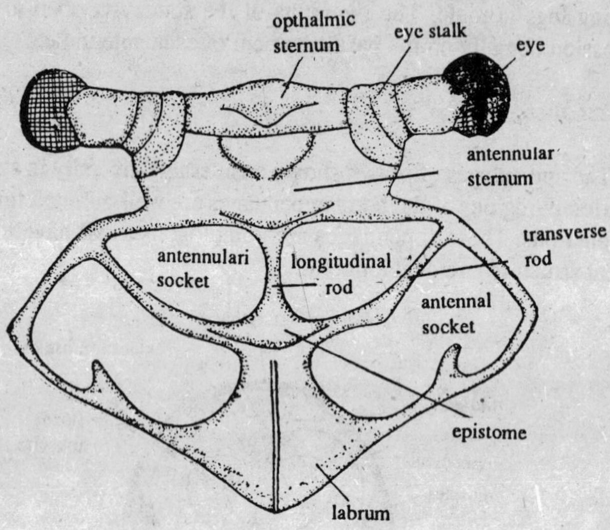

Fig 2.3 *Palaemon*. Front view of preoral region

There are 19 pairs of appendages which can be studies under 3 headings.

I. Cephalic appendages.

II. Thoracic appendages.

III. Abdominal appendages.

I. Cephalic appendages.

As in all arthropods, there are 6 segments of the head. On the first there is a pair of compound eyes. These are not serially homologous with the other appendages, since they arise in a different way. Thus there are

5 pairs of appendages in cephalic region. These are antennules, antennae, mandibles, maxillulae and maxillae.

1. *Antennules.* These are just beneath the eye-stalks. Each consists of a protopodite, which is made up of three podomeses called *precoxa, coxa* and *basis.* The basal podomere, the precoxa is very large and hollow. The cavity of precoxa lodges the *statocyst*, whose opening is situated near its proximal end. From the outer margin of precoxa arises a spiny lobe, the *stylosclerite* . The coxa is short and cylindrical. It bears setae on its inner and anterior borders. The basis is longer than the coxa and lacks setae. It bears a pair of long, slender, many - jointed, tapering processes, the *feelers,* with swollen bases. The inner feeler is unbranched but the outer one is divided into two unequal branches, the inner smaller branch bearing special olfactory setae. It is wrong to regard the two feelers as the two rami of biramous limb, perhaps they represent only the endopodite subdivided.

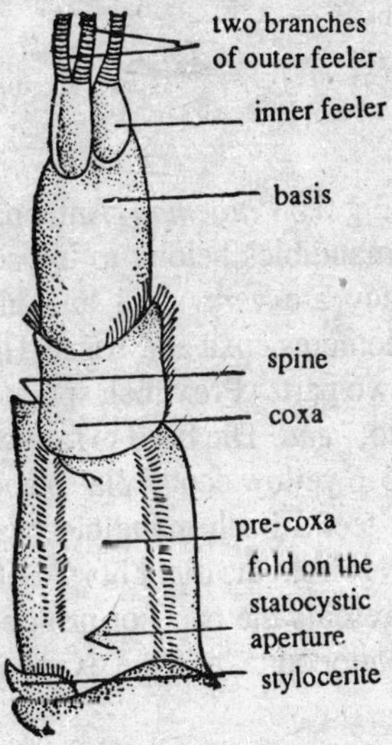

Fig. 2.4 *Palaemon,* Antennule

2. *Antennae.* The antennae belong to segment 3, each with one feeler, situated immediately below and behind the antennules. The protopodite has two podomeres, the coxa and basis. Each one is swollen

at the base as it lodges the renal glands within. The coxa is small in size and bears the renal aperture on its inner margin. The basis bears two rami: a long, slender, many jointed *feeler* with a swollen basis and a broad leaf-like *squama*. Of these the feeler is corresponds to the endopodite and the squama the exopodite. Squama acts as a balancer in swimming, is fringed with setae along its inner broder. Its outer border is smooth terminating in a small spine in front.

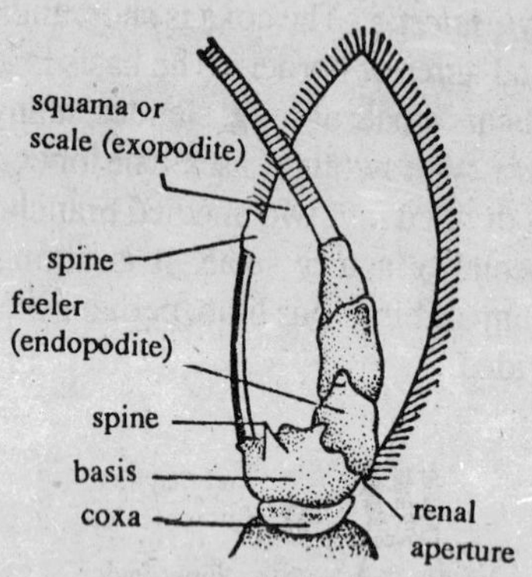

Fig.2.5 *Palaemon* Antenna

3. *Mandibles*. The mandibles belong to the segment 4, lie on the sides of the mouth. They are very stout and highly calcified. The protopodite has two podomeres coxa and basis. The coxa is very large and can be divided into two parts : Proximal, spoon-shaped hollow *apophysis* and the distal solid, *head*. The head is formed of two processes - a *molar process* having 5-6 yellow dental plates and a plate like *incisor process* ending in three teeth. Each mandible bears three segmented *mandibular palp* that arises from the outer border of the head. The basal segment of the palp represents the basis of protopodite, the outer two segments represent the endopodite, and the exopodite is absent.

4. *Maxillulae*. These are small thin, and leaf-like appendages of 5th segment. Coxa and basis are broad, and project inward as *gnathobases* or *jaws* fringed with bristles. Endopodite forms a curved process bifurcated at the apex. The exopodite is absent. The maxillulae halp in the manipulation of food.

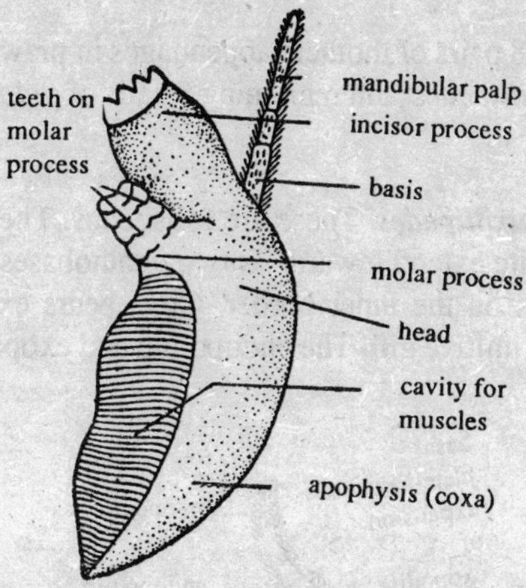

Fig.2.6 Palaemona Mandible

5. *Maxillae*. These are thin, flatenned and delicate leaf-like structure present in the last cephalic segment. Coxa is small and incompletely divided into two representing a fusion of procoxa and coxa. The basis is large and produced into a forked and strongly setose gnathobase, the limbs of the fork are called *endites*. Endopodite is small and unjointed but the exopodite is large, flatenned and fan-like called *scaphognathite*. The movement of scaphognathite creat a water current passing over the gills. It also bears sensory setae along its free margin. The maxillae help in manipulation and respiration.

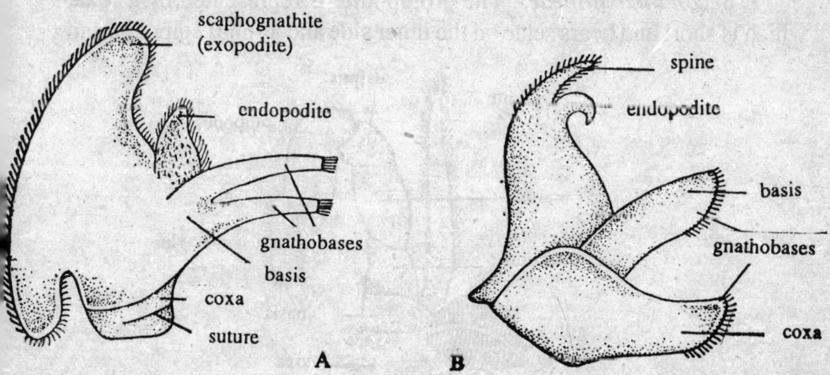

Fig.2.7 *Palaemona* A=Maxilla B = Maxillula

11.Thoracic Appendages

There are 8 pairs of thoracic appendages in prawn. Anterior 3 pairs are called *maxillipedes* and remaining 5 pairs of *palaeopods or walking legs.*

1. *First maxillipedes.* These are foliaceous. The coxa and basis of their protopodite extend inwards as two gnathobases, each bearing two rows of spines on the inner border. Coxa bears externally a bilobed *epipodite* or primitive gill. The endopodite and exopodite are unjointed and bear setae.-

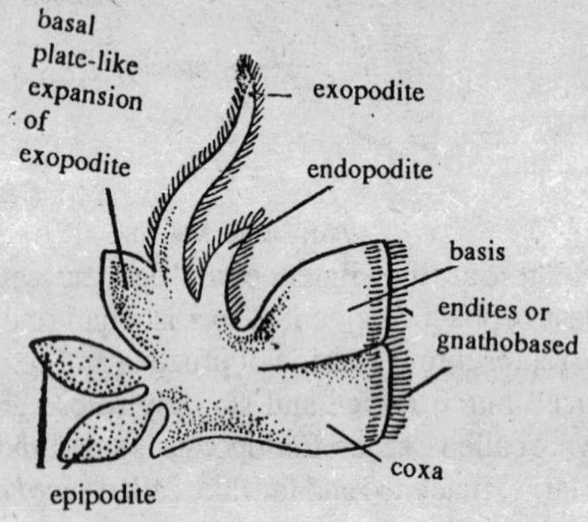

Fig.2.8 *Palaemona* I. Maxillipede

2. *Second maxillipedes.* The protopodite is represented by a coxa, which is short and bears setae on the inner side and a small epipodite and

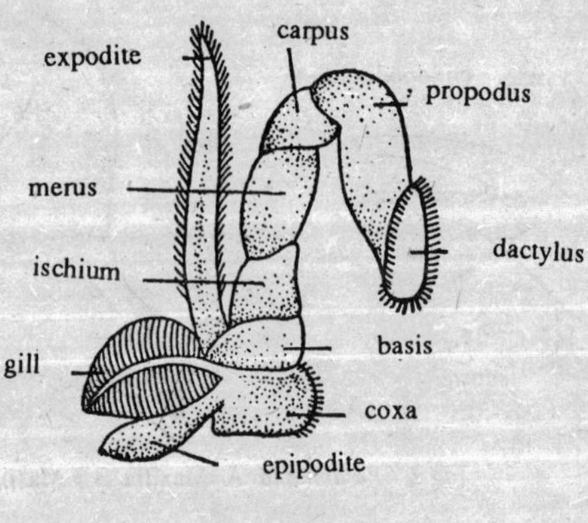

Fig.2.9 *Palaemona* II Maxillipede

a gill on the outer border. Basis is short and is immovably joined with the endopodite which consists of five podomeres: (i) *Ischium*, (ii) *Merus*, (iii) *Carpus*, (iv) *Propodus* and (v) *Dactylus*. Last two are bent backwards and inwards. Exopodite is long, slender, unjointed and is fringed with setae. The functions of second maxillipede are also the same as those of the first maxillipede i.e. to help feeding as well as respiration.

3. *Third maxillipedes.* These are leg-like in appearance but possess same parts as 2nd maxillipedes. The coxa bears an *epipodite* on the outer side and setae along the inner border. The basis carries outer unsegmented *expodite* and an inner three-jointed *endopodite*. The first podomere represents fused ischium and merus, the 2nd carpus and the third represents fused propodus and dactylus.

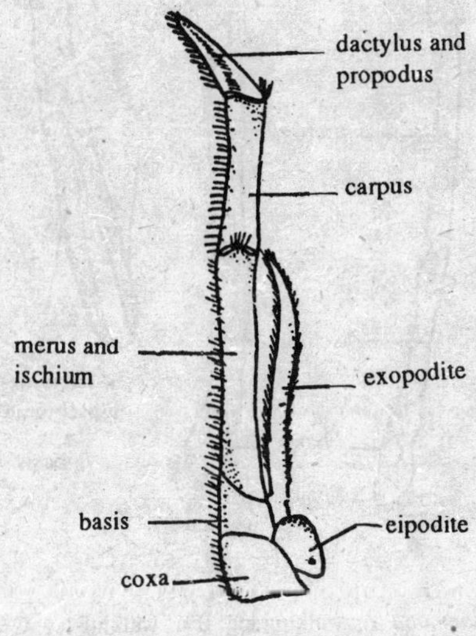

Fig.2.7 *Palaemona* III Maxillipede

4. *Walking legs.* The 5 pairs of walking legs differ from the maxillipedes in the absence of exopodite and epipodites. Thus they have two segmented protopodite and a five segmented endopodite. These 7 segments are arranged in row. The fourth pair of legs represent the typical structure. It contains all the 7 podomeres as *coxa, basis, ischium, merus, carpus, propodus* and *dactylus*.

In the first and second pairs of walking legs the propodus is prolonged distally beyond its articulation with dactylus and thus forms a long forceps-like structure. The structure is called *chela* or *pincer* and the leg is called *chelate leg*. The chelate legs are used to catch the prey and transfer it to the mouth and also serve as the organs of offence and défence. In male the 2nd pairs of chelate legs are larger and more stronger than that of female.

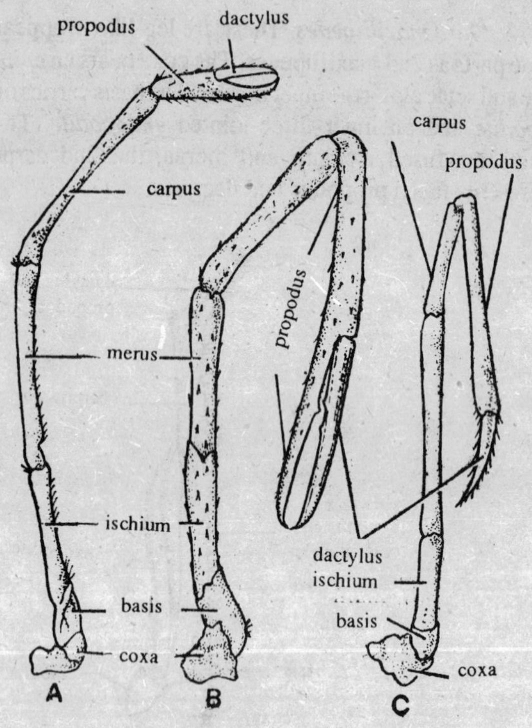

Fig.2.11 *Palaemona* A=walking leg, B=II walking leg, C=III Walking leg.

The third, fourth and fifth pairs of walking legs are typical and non-chelate. In female, each of the third leg bears a crescentic female genital aperture on the inner side of its coxa, while in male, each of the fifth leg bears a slit-like male genital aperture on its inner side of the arthrodial membrane connecting the leg and thorax.

III. Abdominal appendages.

There are 6 pairs of abdominal appendages, These are collectively called *pleopods* or *swimmerets*. These are biramous appendages and help in swimming.

In a typical swimmeret (the 3rd abdominal appendage) the protopodite consists of two podomeres, a short ring-like segment devoid setae known as the coxa and the basis which is cylindrical, bears setae on its surface and carries exopodite and endopodite. The endopodite is a leaf-like structure and gives of a small, rod-like structure, the *appendix interna,* with a knob like head bearing many hook-like processes.

Differences from the typical swimmeret described above, are to be seen in the 1st, 2nd and 6th abdominal appendages of male and 1st and 6th of the female.

First abdominal appendages. In these endopodite is reduced and appendix interna is absent.

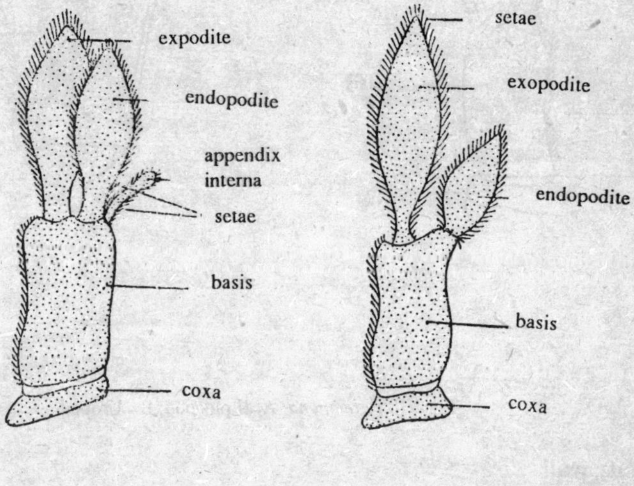

Fig.2.12*Palaemona* A=Typical abdominal appendage, B= I pleopod

Second abdominal appendages of male. These are typical in female but in male the appendix interna gives off a rod-shaped process on its inner side called *appendix masculina.* It is covered with setae.

Sixth abdominal appendages. The sixth pair of abdominal appendages, called *uropods,* are large and lie one on either side of the telson forming *tail-fin* which enables the prawn to take a backward spring in water. The coxa and basis fuse together to form a triangular *sympod,* bearing the oar-shaped endopodite and exopodite. The exopodite is bigger than the endopodite and incompletely divided in the middle by a transverse suture. Their margins, except the outer border of the exopodite, are fringed with numerous seatets

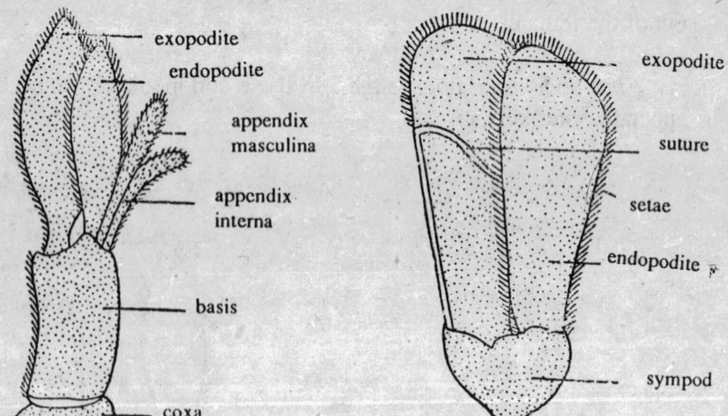

Fig.2.13 *Palaemona* A=II pleopod, B= Uropod

Body wall

The body wall of *Palaemon* consists of an outer layer of *cuticle* a middle single layered *epidermis* and an inner *dermis.*

Table 2.1 Appendages of Prawn

Body Region	*No*	*Name of Appendage*	*Protopodite*	*Endopodite*	*Exopodite*	*Special features*	*Functions*
	1.	Antennules (pre-oral)	Consists of three podomered, (a) Pre-coxa (b) Coxa, and (c) Basis.	Long, jointed feeler with swollen base.	Represented by an outer feeler which is further divided into an inner smaller branch and an outer large branch.	Procoxa has a cavity in which statocyst lodged.	Tactile, is olfactory and equilibrium
	2.	Antennae (Pre-oral)	Two podomeres: (a) Coxa, and (b) Basis.	Represented by a longer feeler having three podomeres.	Flat, broad, leaf like squama or scale.	Both coxa and basis are swollen due to the excretory organ with them. Coxa possesses a renal opening.	Sensory, excretory and balancing.
Cephalic appendages (5)	3.	Mandibles (post-oral)	Coxa large, spoon-shaped and densely calcified, Basis consists of two Parts :- (a) apophysis, and (b) head with molar process and incisor process. Basis is short	Represented by a mandibular palp which is two jointed.	Absent.	Coxa-teeth are located at the side of the mouth.	Mastication of food

Body Region	No	Name of Appendage	Protopodite	Endopodite	Exopodite	Special features	Functions
	4.	Maxillulae (Post-oral)	Coxa and basis both are leaf-like and project as gna thobases covered with pointed spines.	Bifurcated process on the outer side of the basis.	Absent.	Manipulation of food.	
	5.	Maxillae (post-oral)	Coxa is incompletely divided while basis forms a bifurcated gnathobasis.	Reduced to a small process which is narrow and is forwardly direct.	Flat, fan-shaped scaphognathite	Scaphognathite produces currents of water over gills.	Respiration and manipulation of food.
	6.	1st Maxillipede	Coxa and basis project like jaws	Endopodite is represented by a small, unjointed appendage.	Exopodite is also unjointed and a plate-like expansion.	Similar to a leaf in shape	To hold food and tactile.
	7.	2nd Maxillipede.	Protopodite is represented by coxa which is short and bears seate on the inner side, and a small epipodite and gill on the outer border. Basis is short and is immovably joined with the endopodite.	It consists of 5 podomeres: 1. Ischium, 2. Merus. 3. Carpus. 4. propodus. 5. Dactylus.	Represented by long, slender, unjointed and fringed setae.		To help in feeding.

Body Region	No	Name of Appendage	Protopodite	Endopodite	Exopodite	Special features	Functions
Thoracic appendages (8)	8.	3rd Maxillipede.	Coxa is short and possesses epipodite on the inner side. Basis is short and supports exopodite and endopodite.	It consists of 3 joints:- 1. Fused ischium and merus. 2. Carpus. 3. Fused propodus and dactylus.	Unjointed, long, slender and thickly covered with setae.	Leg-like in appearance.	To hold the food in position and to help in feeding.
	9.	Ist Chelate leg.	Coxa is short and ring-like, while basis is triangular.	5. distinct podomeres, i.e., Ischium, merus, carpus, propodus and dactylus	Absent.	Chela or pincer is Present.	Catches and transfers the prey to mouth.
	10.	2nd Chelate leg.	" "	" "	" "	" "	Organ of offence and defence. Large in males.
	11.	3rd Non-chelate leg.	" "	Propodus and dactylus do not form chela.	" "	Chela absent. In female, coxa bears a female genital aperture	Walking
Thoracic appendages.	12.	4th Non-chelate leg.	"	"	"	Chela absent.	Walking.
	13.	5th walking leg.	"	"	"	In male, there is a male genital aper-	Walking

Body Region	*No*	*Name of Appendage*	*Protopodite*	*Endopodite*	*Exopodite*	*Special features*	*Functions*
						ture between the leg and thorax.	
Abdominal appendages (6)	14.	Ist pleopod or swimmeret	Ring-like coxa and cylindrical basis. Basis possesses two rami.	Small, leaf-like, possessing setae.	Large, leaf-like.		Swimming.
	15.	2nd pleopod.	"	In female, it helps to carry the eggs with 3rd, 4th and 5th, appendix interna.	"	Appendix masculina is also present between endopodite and appendix. interna.	Appendix interna is present on the inner side on the endopodite as a short, slender, slightly-curved rod. Appendix interna is interlocked with its fellow of the opposite side to form a basket-like structure.
	16.	3rd pleopod.	"	"			
	17.	4th pleopod.	"	"	"	"	
	18.	5th pleopod.	"	"	"	"	
	19.	Uropod or 6th swimmeret.	Coxa and basis fused to form a broad plate or sympod.	Flat, oval plate and provided with setae.	Large and divided into two pairs by a transverse suture.	With the telson form tail.	Swimming

1. *Cuticle.* The cuticle of the body wall has become transformed into the exoskeleton which is thick and rigid at some places and thin and membranous at others. Cuticle is composed of two layers: (i) the outer layer is a thin non- chitinous *epicuticle.* It is yellow hyaline layer. It is further made up of two layers. Outer extremely thin made up of lipids and inner relatively thick protein layer. The protein layer is hardered and pigmented by its combination with an oxidised phenol. The epicuticle is secreted by the tegumental gland cells present is dermis. (ii) The *endocuticle* contains layers of chitin and has three successive layers: the outermost layer is a thin pigmented layer, below it is thick calcified layer and under it is an uncalcified layer. The pigmented layer is dark due to protein deposists and it becomes hardened by a process of tanning or sclerotization. In the calcified layer are deposits of carbonates and phosphates of calcium which make the integument hard. The chitin is an acetate of a polysaccharide containing glycosamine secreted by the underlying epidermis. The cuticle is relatively impermeable, except where it is thin and allows the passage of gases or absorption of water.

The seate and spines are the outgrowths of cuticle and have the same structure.

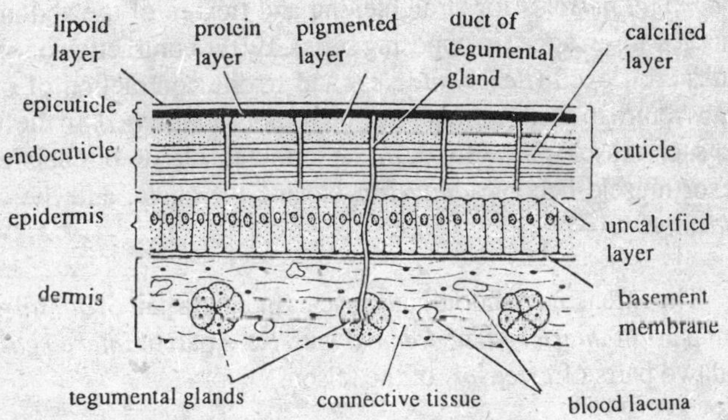

Fig.2.14 *Palaemona* V.S. of Skin

2. *Epidermis.* It comprises a single layer of glandular colummar epithelium resting on a thin basement membrance which is secreted by the underlying dermis. The epidermis is often called *chitogenous* epithe-

lium as it secretes chitin.

3. *Dermis.* The dermis is thin and is made up of loose connective tissue beast with blood lacunae. It contains three types of tegumental glands that open out by fine ductules passing through overlying epidermis and the cuticle.

Body cavity

In *Palaemon* the body cavity is mostly occupied by musculature and organs (having interspaces containing blood) and all these together form a body cavity or haemocoel. However, true coelom is restricted to kidneys and gonads.

Musculature

The bulk of the body of the *Palaemon* consists mainly of muscles which are enclosed within the exoskeleton. There muscles are voluntary or striated type. The unstriated muscles are found in the organs such as blood vascular system and digestive system. There are two sets of muscles in prawn, the *appendicular muscles* that are segmentally and the *abdominal muscles* for straightening and flexure of the abdomen. The muscles are arranges in opposing series. By the contraction of one set the limbs are flexed (*flexor muscles*) and by the contraction of other, the limbs return to their normal position (*extensor muscles*). In the abdomen too such muscles are found. In the terminal sclerite of chelate leg, the flexor muscle become *adductor*, closing the chela, and the extensor, *abductor*.

The extensor of abdominal muscles include a pair of *dorsal extensor*, four pairs of *intertergal abdominal muscles*, a pair of *intertergal muscles* and two pairs of *extensor* of the telson.

The flexors of abdominal muscles include 5 pairs of main flexor (first 5-segments), four pairs of *accessory flexor* muscles (first 4 segments), a pair of flexor (between first abdominal segment and thorax), two pairs of *lateral thoraco- abdominal* muscles, two pairs of flexor muscles of telson and a pair of *ventral superficial flexor* muscles.

All the appendages have their own muscles.

Endoskeleton

Palaemon has an endoskeleton in the cephalothorax formed of a series of ingrowths of the exoskeleton called *apodemes* or *endophragms*. These ingrowths are connected together to form a regular framework that is known as *endophragmal skeleton*.

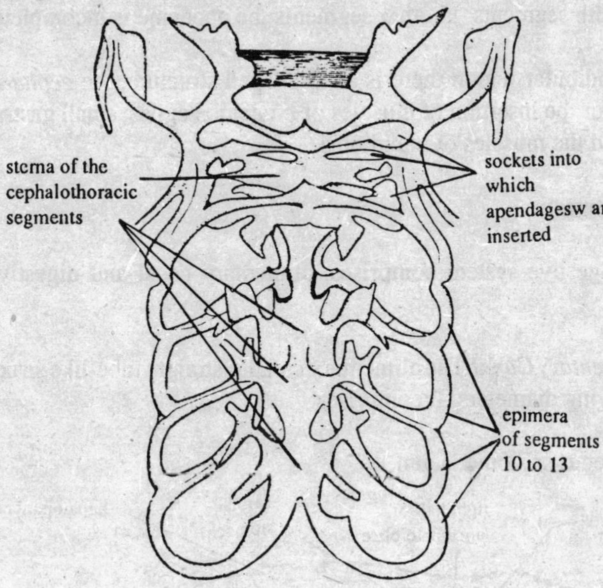

Fig.2.15 *Palaemona* The sterna and endophragmal skeleton (dorsal view)

The endophragmel skeleton is well developed in the segments 11-13, where, in each segment, it consists of an outgrowth from the antero-lateral end of epimeron of the same sternal segment called *endopleurite posterior*.

The endopleurite posterior meets an ingrowth from the posterior-lateral end of the epimeron of the segment in front, called *endopleurite anterior*. These in turn meet respective ingrowths of the sternum of the same segment an *endosternite posterior* and an *endosternite anterior*. From the mesosternite, on each side arises an upwardly directed 'Y' shaped plate. The inner limb of the Y is called *mesophragm* and outer *paraphragm*. These apodemes provide for attachment of abdominal flexor muscles while other apodemes are for the attachment of appendicular muscles. A typical apodeme with complete number of endosternites, endopleurites and Y shaped structure occurs only between 11th, 12th and 13th segments. In other segments the apodeme is incomplete.

In mandibular region there is an H-shaped structure, the *cephalic apedome,* for the insertion of muscles of branchiostegites, small gastric muscles and the muscles of mandibles.

Digestive System

The digestive system comprises, alimentary canal and digestive glands.

1.*Alimentary Canal.*The alimentary canal is straight tube-like structure of varying diameters. It comprises:

1. Foregut or stomodaeum.

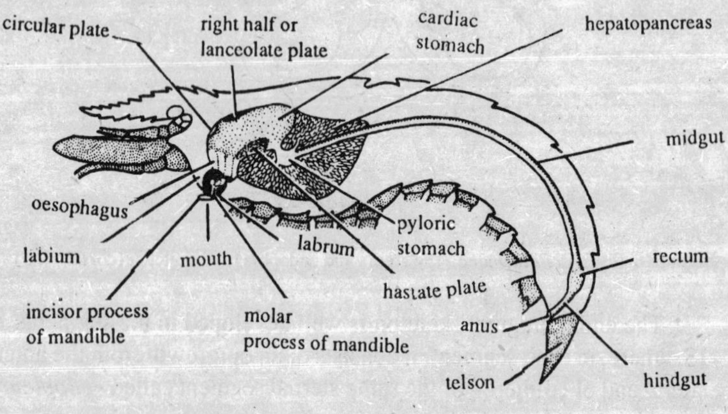

Fig.2.16 *Palaemona* Alimentary canal in lateral view

2.Midgut or mesenteron

3. Hindgut or proctodaeum.

1. *Foregut*. The foregut comprises the mouth, buccal cavity, oesophagus and stomach. Foregut is ectodermal in origin hanace it is internally lined by ectoderm and cuticle.

(i) *Mouth*. The mouth is large slit-like apertuse situated on the ventral side of the head below the anterior end. It is bounded anteriorly by a shield shaped *labrum* or upper lip. Laterally by the incisor processes of *mandibles* and posteriosly by a thin *labium* or lower lip which is bilobed. Mouth leads into the buccal cavity.

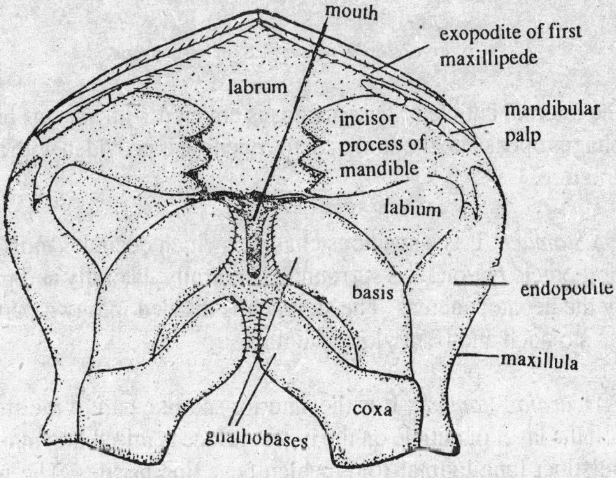

Fig. 2.17 Palaemon ventral view of oral region

(ii) *Buccal cavity*. The *buccal cavity* is an antero-posteriorly compressed small chamber with irregularly folded thick chitnous lining. The *molar processes* of mandibles project into it and lie opposite each other.

(iii) *Oesophagus.* The oesophagus is a short, wide, vertical tube joined with the buccal cavity and stomach. Its inner wall is muscular and is produced into four longitudinal folds: an anterior fold which is short, aposterior fold and two lateral folds, which are large and more prominent.

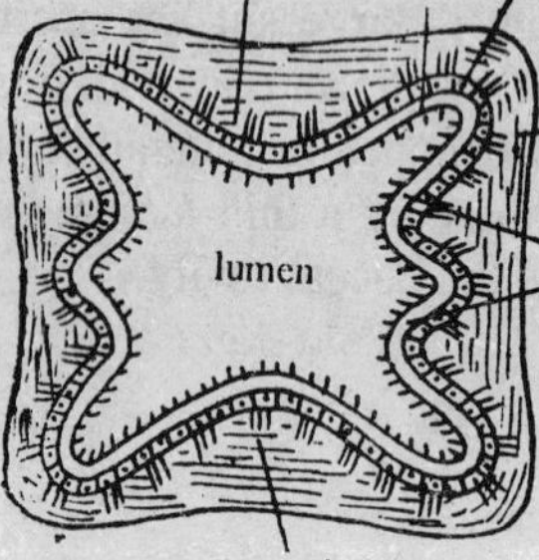

Fig. 2.18 Palaemon T.S. of oesophagus

The internal cuticle of oesophagus is covered with various bristels. Oesophagus opens into the stomach; the opening is provided with a valve like structure.

(iv) *Stomach.* It is a spacious chamber which occupies most of the cephalothoracic region. It is surrounded ventrally, laterally and posteriorly by the hepatopancreas. The stomach is divided into two parts. (A) Cardiac stomach and (B) Pyloric stomach.

(A) *Cardiac stomach.* It is the anterior sac-like part of the stomach in which the layer of cuticle on the inner surface is raised into many fine and indistinct longitudinal folds which bear fine bristles. The wall of stomach is supported by a number of cuticular plates. These plates are as follows:

(a) *Circular plate* is present in front of oesophageal aperture forming its anterior wall.

(b) *Lanceolate plate* is embedded mesially in the roof of the stomach just behind the circular plate

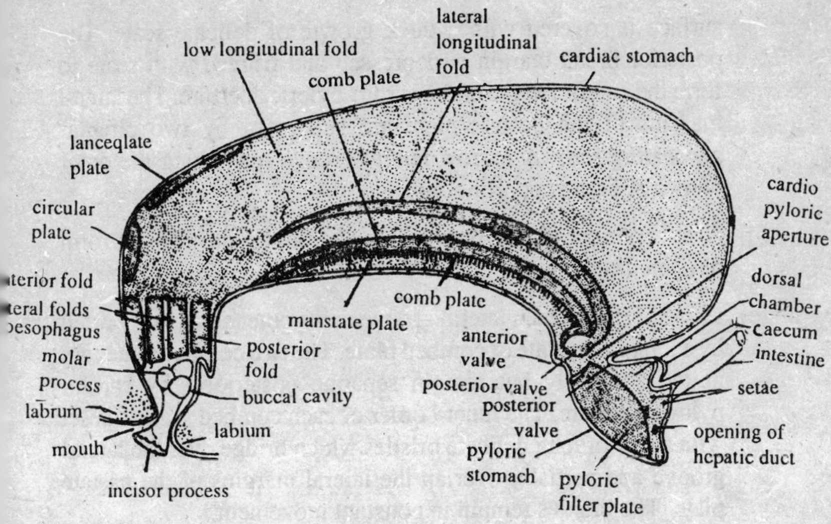

Fig. 2.19 Palaemon V.L.S.of Foregut

(c) *Hastate plate* is a triangular plate embedded in the middle of floor of stomach. It is raised in the middle with sloping sides. Its upper

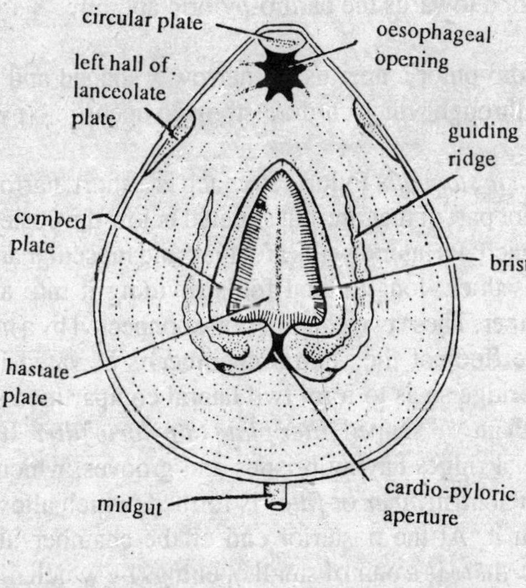

Fig. 2.20 Palaemon Floor of cardiac stomach

surface is covered with a thick growth of delicate seate. The posterior broad margin is depressed and fringed with seate to form the anterior valve of the cardio-pyloric aperture. The lateral margins of the hastate plate are supported by two circular *supporting rods.* A narrow lateral groove runs along the outer border of each supporting rod.

(d) *Groove plates* form the floor of lateral grooves and are in the form of open drain pipes.

(e) *Combed Plate.* Each lateral groove on its outer side is supported by a long ridge plate or combed plate. The two combed plates are united anteriorly but remain separate posteriorly by cardio-pyloric aperture. The inner border of each combed plate is beset with a row of long delicate bristles which bridge over the lateral groove and partially overlap the lateral margins of the hastate plate. The bristles remain in constant movement.

Just outside the combed plates, the wall of cardiac stomach is folded on each lateral side to form *guiding ridges* or *lateral longitudinal folds.* These ridges are low in front but gradually increase in high behind, but posteriorly the high walls of these folds bend inwards over the cardio-pyloric aperture, forming lateral valves. Thus these valves guard the passage of food towards the cardio-pyloric aperture.

The cardio-phloric apertuse is narrow x shaped and is guarded by four valves, through which cardiac stomach opens into pyloric stomach

(B) *Pyloric stomach.* Pyloric stomach is a short, narrow chamber. It is the posterior part of the fore-gut situated below the posterior end of the cardiac stomach. Its lateral walls are thick and muscular and project into the lumen as thick, longitudinal folds dividing it into a dorsal and a ventral chamber. These two chambers are connected by a narrow vertical passage. The floor of the ventral chamber is elevated into a median longitudinal ridge so as to form two lateral compartments. The floor is provided with an '^' shaped *filter plate* or *pyloric filter*. It bears a series of longitudinal ridges having bristles and grooves, which are felt-like. Thus, an efficient *strainer* or *filter* is formed, which allows only liquid food through it. At the posterior end of the chamber just behind the pyloric filter, there is a pair of small openings by which digestive juices from the hepatopancreatic duct are poured into the pyloric stomach.

These openings are provided by posteriorly-directed setae, which act as valves to control the direction of the flow of the secretion.

The dorsal chamber gives out a small blind caecum dorsally and opens into the mid-gut. The junction of dorsal chamber and midgut is guarded by one median dorsal and two lateral groups of elongated setae. These setae control the flow of food material.

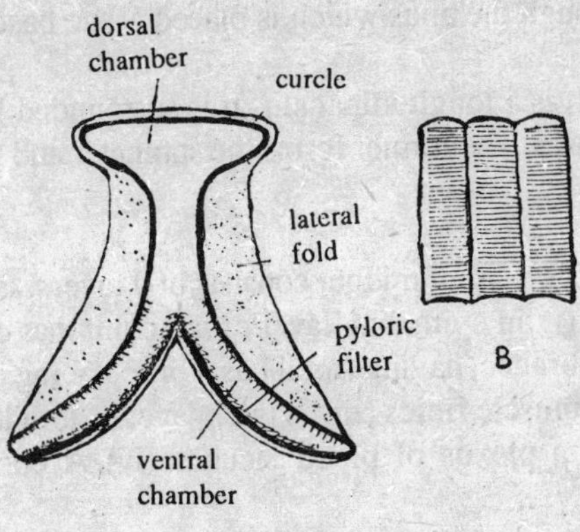

Fig. 2.21 Palaemon A=T.S. pyloric stomach B= A part of pyloric stomach

(v) *Midgut or mesenteron.* The midgut is a long, narrow, stright and slender tube running back along the median line above the mass of the

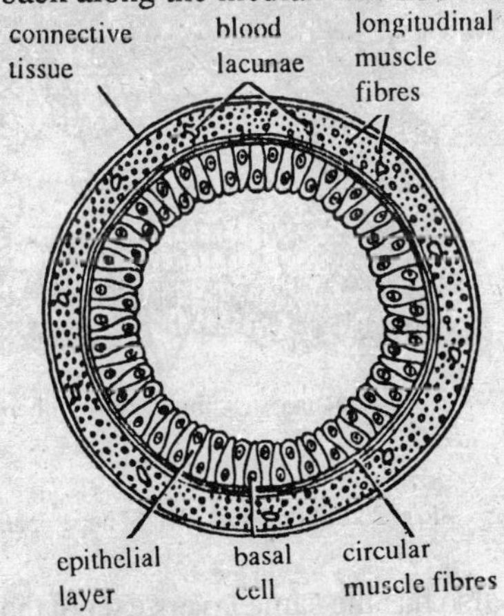

Fig. 2.22 Palaemon T.S. of mesenteron

ventral abdominal muscles upto the sixth abdominal segment. The internal lining of midgut is formed by epithelium which in the posterior part is thrown into many longitudinal folds thus greatly reducing its lumen.

(vi) *Hindgut or proctodaeum.* It is the last part of alimentary canal which is internally lined with cuticle. Its anterior enlarged muscular part is called the *intestinal bulb* and the posterior less wide part is the *rectum.* It opens out through the anus which is placed at the base of the telson.

(vii) *Anus.* It is a longitudinal slit. It is surrounded by circular and radial muscle fibres. The former forms the sphincter and the later acts as dilator.

Histology. The wall of midgut consists of 4 layers. Beginning from inside, these are (i) an epithelial layer of tall columnar cells resting on a basement membrance, having small basal or replacing cells (ii) A thin layer of circular muscle fibres (iii) A thick layer of longitudinal muscle fibres enclosing a plexus of blood lacunae (iv) A connective tissue sheath.

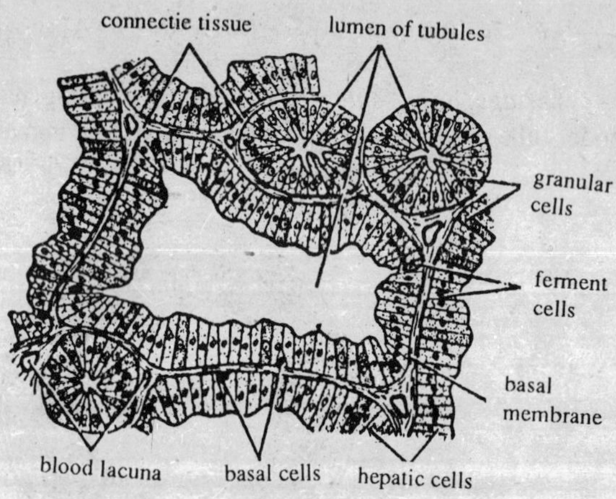

Fig. 2.23 Palaemon T.S. of hepatopancreas

The rectum also has the same layers except that its epithelial lining is covered by cuticle and there is an additional layer of longitudinal muscle fibres inside the circular muscle fibres.

II *Digestive gland.* The *hepatopancreas* is a large and physiologically very important gland. It consists of a pair of orange-red coloured lobes lying close to each other surrounding the cardiac stomach and has a massive appearance. It is a racemose gland which arises as a pair of tubular outgrowths from the midgut. These outgrowth grow, divide and redivide, thus forming a compact network which is held together by connective tissue. The wall of each tubule forming the gland is made up of four layers: the inner most layer of epithelial cells, a thin basement membrane on which epithelial cells rest and the outer most layer or *tunica propria* formed of a network of connective tissue and muscle fibres. The epithelial layer has four kinds of cells (i) *Hepatic cells* which are columnar shape and with their cytoplasm filled with fat droplets (ii) granular cells (iii) irregularly scattered forment cells and (iv) basal or replacing cells.

The hepatopancreas perform following functions:

1. It secretes ensyme for digestion of carbobydrates, fats and proteins.

2. It serves to store glycogen, calcium and fat.

3. It helps in absorption of digested food material.

Food and feeding *Palaemon* is omnivorous, feeding upon aquatic weeds, algae, moss, insects and debris of bottom. It feeds at night. The chelate legs, aided by the third maxillipedes, capture and convey the food to the mouth. The coxae of the second maxillipedes hold the food, while the incisor processes of mandibles cut it into smaller pieces which are swallowed with the help of maxillipedes, maxillulae and maxillae. Inside the buccal cavity the molar processes of mandibles masticate the food, which is then conveyed to the cardiac stomach through the oesophagus.

Digestion. The digestion of food starts in the cardiac stomach where the hepatopancreatic juice from the pyloric stomach ascends up and mixes well with the food. Due to the alternate expansion and contraction of stomach wall the food is well churned. While passing over the hastate plate the food particles are still reduced to minor pieces so that the food is converted into a paste. This paste is filtered through the bristles of combed plate and is carried to the pyloric stomach. The food in the semi-liquid and semi-digestid condition is carried into the ventral chamber of pyloric stomach. The digested and liquified food is filtered through the pyloric filter and it again enters the hepatopancreas through the hepato-pancreatic ducts and then becomes absorbed. The hepatopancreas served for digestion as well as for absorption. The residual food material, consisting of undigested particles, passes on to the midgut and the remaining digested food is absorbed in the midgut. The undigested faecal matter pass on to the posterior side of the hindgut and ultimately expelled through the anus.

Respiratory system

The respiratory organs is prawn are developed and are located in the gill chambers, between the branchiostergite and the thorax. These are :

1. Branchiostergite or gill cover.
2. Three pairs of Epipodites.
3. Eight pairs of Gills or brnachiae.

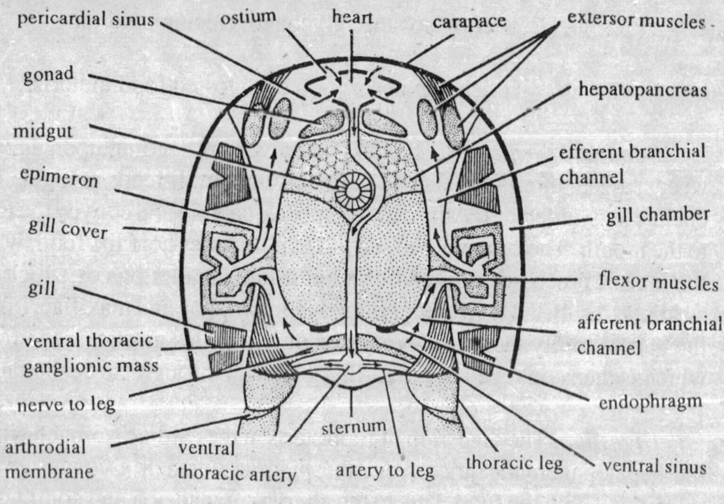

Fig. 2.24 *Palaemon* Section through cephalsthorax showing gill chamber

1. *Branchiostergite*. The inner lining of branchiostergites is thin, membranous and highly vascular containing minute blood lacunae. These are constantly bathed in fresh water, thus forming the respiratory surface through which exchange of gases takes place between blood and water.

2. *Epipodites*. These are thin, leaf-like processes, one attached to each maxillipede. These epipodites are full of blood lacunae and, infact, constitute a primitive type of gills. The epipodites lie in the anterior part of the gill-chamber below the scaphognathite. Exchange of gases between blood and lacunae takes place between water and blood in epipodites.

3. *Gills*. There are 8 pairs of gills. These arise from the modification of epipodites and according to their position, they be distinguished into three types:

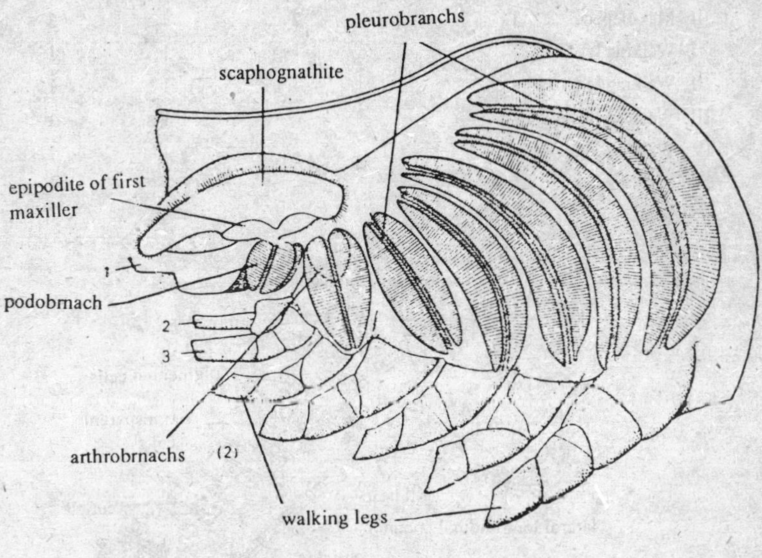

Fig. 2.25 *Palaemon* Gill Chamber exposed to show arrangement of gills

(i) *Podobranch*. When a gill is attached to the coxa of an appendage, it is called podobranch or foot-gill. In *Palaemon* one podobranch is carried by the coxa of each 2nd maxillipede.

(ii) *Arthrobranch*. When a gill is attached to the arthrodial membrance connecting the appendage to the thorax it is called arthrobranch or joint gill. Each 3rd maxillipede bears two arthrobranches.

(iii) *Pleurobranch*. When a gill is attached to the lateral wall of the segment bearing limb, it is called pleurobranch or side gill. The last five gills on each side are pleurobranch attached to the lateral wall of the thoracic segments bearing five walking legs.

Branchial formula. The number and position of epipodites and gills can be illustrated by the branchial formula which is given in the following table.

Appendage	*Epipodite*	*Podobranch*	*Arthrobranch*	*Pleurobranch*	*Total*
I Maxillepede	1	-	-	-	1
II Maxillepede	1	1	-	-	2
III Maxillepede	1	-	2	-	3
I Walking leg	-	-	-	1	1
II Walking leg	-	-	-	1	1
III Walking leg	-	-	-	1	1
IV Walking leg	-	-	-	1	1
V Walking leg	-	-	-	1	1
Total	3	1	2	5	11

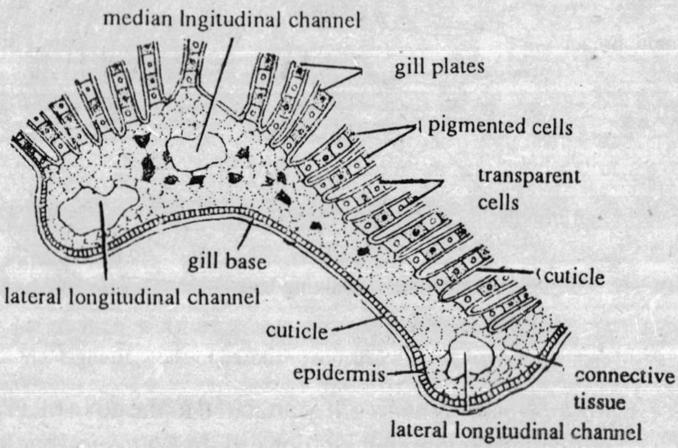

Fig. 2.26 *Palaemon* T.S. of gill

Structure of gill. The gills are almost crescentic or semilunar in shape. Each gill consists of a *central axis* or base from which arise two rows of rhomboidal *gill-plates or lamellae*. Leaving a deep groove called *median longitudinal groove* between the two rows, such a gill is known as *phylobranch*. The gill plates found in the middle of gills are large and gradually become smaller toward both the ends. In the middle of gill arises a small structure called *gill- root* by which it is attached to the thorax. Nerves and blood vessels enter or leave the gill through the gill roots.

The bease of the gill appears triangular in a cross section. It is made up of loose connective tissue which is bounded by epidermis and cuticle. There is a row of epithelial cells in each gill-plates, outside which is a thin layer of cuticle. The cells are of two types *viz. pigmented* and *transparent* , alternating with each other.

Circulation of blood in gills. The gills are profusely supplied with

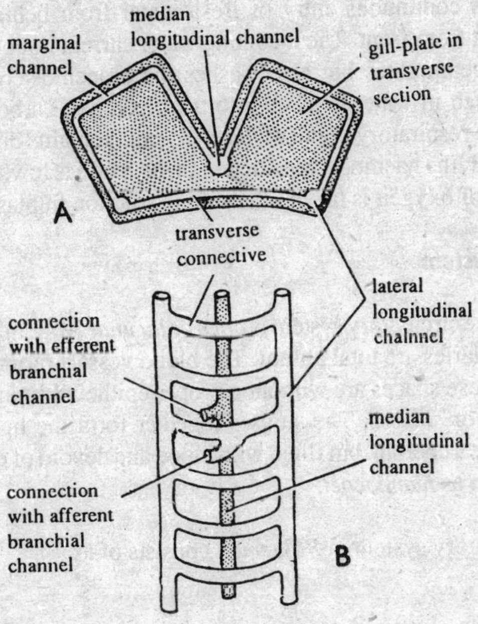

Fig. 2.27 *Palaemon* A=Blood supply in gill plates B= Blood channel in gill

blood. Three blood vessels run vertically in each axis - two *lateral longitudinal* channals along the two lateral margins and a *median longitudinal* channel in the middle. The lateral longitudinal channels are connected with each other by *transverse connectives,* one of which, near the gill - root, communicates with the *affarent branchial* channel and receives blood from it. Each lateral channel gives a delicate marginal channel into each gill-plate which runs along the outer margin of the gill-plate and finally opens into the median longitudinal channel. Blood from the afferent branchial channel passes into the transverse connective from where it goes into the lateral longitudinal channel, then through the marginal channel into the median longitudinal channel which finally pours the blood, at the gill-root, into the *efferent branchial* channel by a short connection.

Mechanism of respiration. The exopodite of maxilla, called scaphognathite, is so situated that its anterior end penetrates into gill chamber and the posterior end lies outside. The gill chamber itself is covered dorsally and laterally but is open ventrally. The vibrating action of the scaphognathite bales out water from the anterior end of the chamber and simultaneously pushed a water current movement of the baler results in continuous entry of fresh water from behind and its subsequent exit from front. The incoming water current not only passes over the gills, epipodites but also washes the branchiostergites. The oxygen dissolved in water is taken into the blood and carbon dioxide passes out. The respiratory pigment, *haemocyanin* remains dissolved in blood. However, the haemocyanin content in the blood is low, therefore, the major part of oxygen is taken is as simple solution in plasma.

Circulatory System

In prawn the circulatory system is *open* or *lacunar type* , in which the veins and capillaries are total absent. The blood vessels (arteries) open into spaces. These spaces are without a proper epithelial lining and are called *lacunae* or *sinuses.* The sinuses together form the body-cavity which looks like a coelom, but filled with blood and devoid of epitheliall lining, is known as *haemocoel.*

The circulatory system in *Palaemon* consists of :

1. Pericardium
2. Heart
3. Arteries
4. Sinuses
5. Blood channels
6. Blood.

1. *Pericardium.* It is a wide thin-walled chamber around the heart and remains filled with *haemocoelomic fluid.* Its floor is in the form of a thin horizontal septum which lies just above the gonad and hepatopancreas and is attached to the dorsal body wall along its anterior and posterior aspects. The septum is attached to the lateral walls of the thorax along the lateral aspect. The pericardial cavity pumps blood into the heart.

2. *Heart.* The heart is a triangular, muscular structure with its broader base facing posteriosly and the apex in front. It is found in the pericardial sinus in the median dorsal part of thorax. A median longitudianl strand, the *cardiopyloric strand,* extending from the apex of the heart to the dorsal wall of the pyloric stomach, and two lateral strands, stratching from the lateral angle of the heart to the body wall, support the

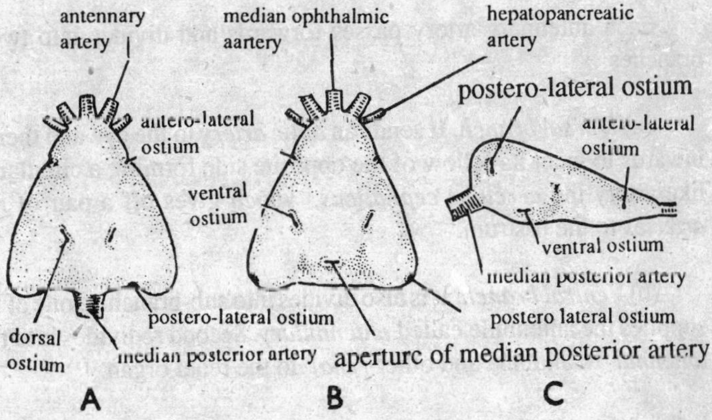

Fig. 2.28 *Palaemon* Heart A=Dorsal view, B= ventral view C= lateral view

heart in the pericardium. The walls of the heart are very thick, muscular and extremely elastic. There are 5 pairs of valvular openings, the *ostia* in the wall of heart. The first pair of ostia is situated in the mid-dorsal region. The second pair is found mid-ventrally. The third pair lies on the posterior border. The fourth pair lie anterio-laterally and the fifth pair posterio-laterally. The cavity of heart is not continuous but is traversed by a large number of interlacking muscle fibres,

3. *Arteries*. The arteries are thick-walled muscular tubes, which arise from the heart and supply blood to different organs of the body. The main arteries are following:

(i) *Median ophthalmic artery*. It arises as a single median cephalic or ophthalmic artery from the apex of the heart. It runs forward to the head region in the mid-dorsal line above the renal sac. Here it joins the two antennary arteries.

(ii) *Antennary arteries*. A pair of them originate from the apex of the heart one from either side of the median ophthalmic. Each runs obliquely forward along the outer border of the mandilbuar muscle. During its course it gives out the following branches:

(a) *Pericardial branch* to the pericardium,
(b) *Gastric branch* to the cardiac stomach, and
(c) *Mandibular branch* to the mandibular muscles.

Each antennary artery passes forwards and divides into two sub-branches:

(a) *Dorsal branch*. It sends an *optic artery* to the eye and then bend inwards to meet its fellow of the opposite side forming a circular loop-like artery the *cerculus cephalicus,* which gives off a pair of *rostral arteries* to the rostrum.

(b) *Ventral branch*. It is also divides into sub-branches, one of which supplies the antennule called *antennulary*. Second redivides into two an *antennal* to antenna and other *renal* to the renal organ.

(iii) *Hepatic arteries*. These are also paired arteries, which arise from the heart on the ventro-lateral sides of the roots of antennary arteries. They ramify into the hepatopancreas.

(iv) *Median posterior artery*. It is a small but stout artery which arises from the posterio-ventral surface of the heart and runs backwards. It immediately bifurcates into two branches.

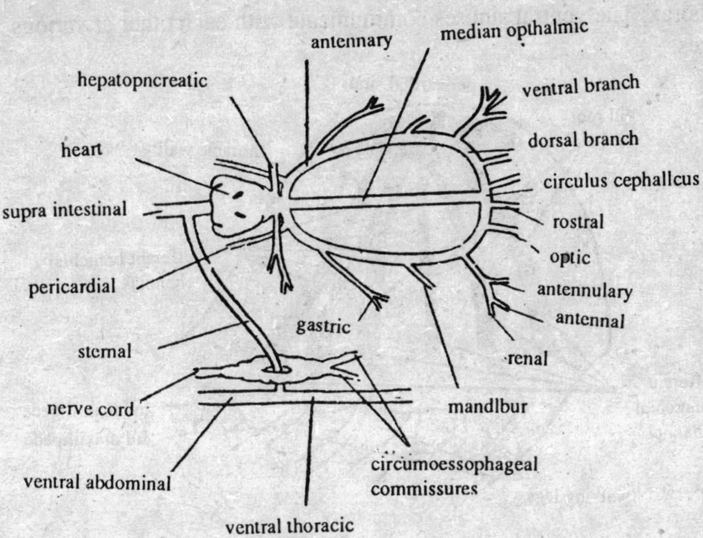

Fig. 2.29 *Palaemon* **Showing heart and arteries**

(a) *Supra-intestinal artery*. The supra-intestinal artery passes straight backwards along the dorsal surface of the intestine extending up to the hind-gut, while it divides into two branches supplying blood to the intestine and the dorsal abdominal muscles.

(b)*Sternal artery*. The *sternal artery* is the stoutest of all others. It runs downards and forwards through the hepatopancreas and after passing through an aperture in the middle of the ventral thoracic ganglionic mass to reach the ventral side, it gets divided into two branches- The *ventral thoracic branch* runs anteriorly up to the mouth and supplies the sternal region, first three pairs of walking legs, maxillae, and maxillulae, etc. The *ventral abdominal branch* runs posteriorly upto the anus and supplies blood to the ventral abdominal region, last two pairs of legs, pleopods and hind gut, etc.

4. *Blood Sinuses*. The various arteries, on reaching the organs, branch repeatedly into minute branches which do not form capillaries but end into wide spaces, *blood sinuses* or *blood lacunae*. All the blood lacunae of the body collectively form a pair of elongated but ill-defined *ventral sinuses*. These lie below the hepatopancreas and above the floor

of thorax. The ventral sinuses communicate with each other at various places.

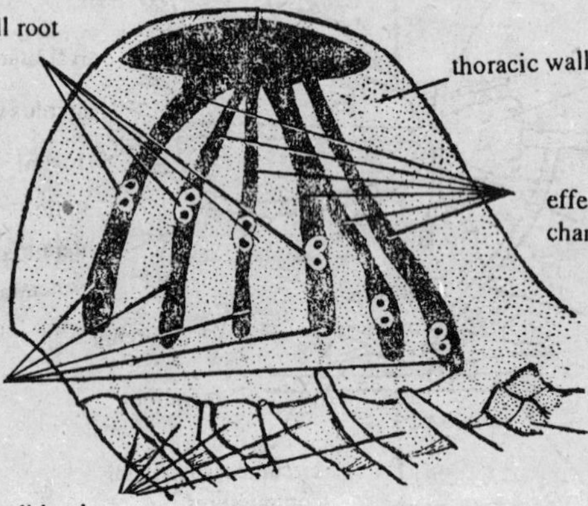

Fig. 2.30 *Palaemon* Branchial channels

5. *Blood Channels.* In *Palaemon*, there are 6 pairs of *affarent branchial channels* and 6 pairs of *efferent branchial channels*. These channels are lacunar in nature. The afferent branchial channels run along the inner surface of the thoracic wall and reach the gills through the gill-roots. Here each channel opens into the transverse connective. The first afferent channel takes the blood to the podobranch and two arthrobranches. Remaining five channels supply the blood to the five pleurobranches. The blood is oxygnated while passing through the gills and this blood is carried to the pericardium by efferent branchail channels.

Course of blood circulation. It can be represented diagrammatically shown as below.

5. *Blood.* The blood is a thin, watery, almost colourless liquid containing floating colourless *amoebocytes*. The colour of the blood is faintly bluish, when it is oxygenated. The colour is due to the presence of a respiratory pigment i.e. *haemocyanin*. It is a compound of copper and protein. There is also a lipochrome pigment called *zoonerythin* in the blood. The blood has the power of coagulation when it comes in contact with air.

Excretory System

The excretory system of adult *Palaemon* consists of a pair of *antennary* or *green glands* and a median sac termed as *renal sac*.

1. *Antennary or green glands.* The antennany glands are opaque white organs of the size of a pea-seed, are enclosed inside the coxae of each antenna. Each gland consists of 3 parts:

(i) End Sac

(ii) Labyrinth

(iii) Bladder.

(i) *End sac.* It is a small bean-shaped structute lying in between the lebyrinth and bladder. Its wall is produced internally into folds and

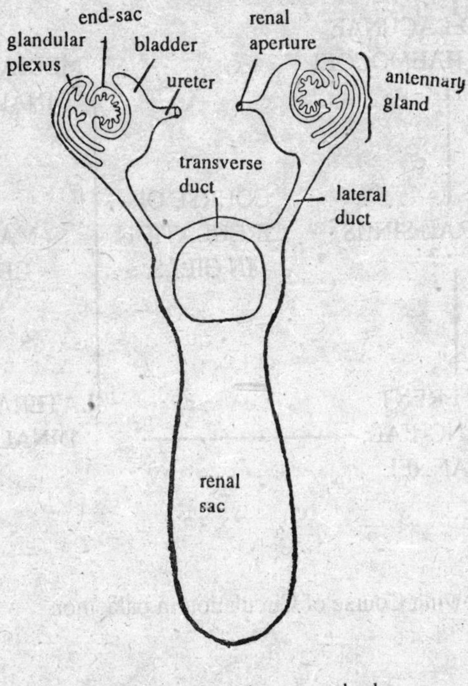

Fig. 2.31 *Palaemon* Antennary gland

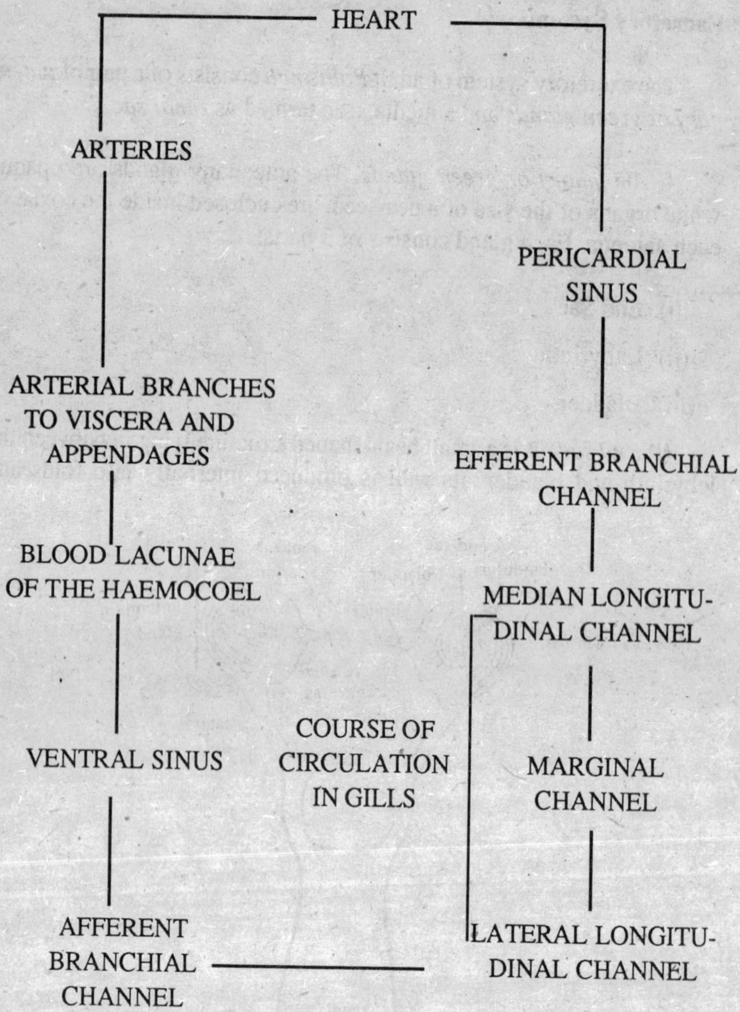

Chart Showing Course of Circulation in palaemon

consists of two layers, outer thick layer of connective tissue with a number of small blood lacunae, and an inner thin layer of excretory epithelium. The cells are large with finely granular cytoplasm and rounded nuclei. The cavity of end-sac contains a large blood lacunae and communicates with the labyrinth by a single aperture.

(ii) *Labyriuth*. The *labyrinth* or *glandular plexus* is very much larger in size then the end sac and lies on its outer side. It is a mass of highly convoluted and extensively branched excretory tubules. Which open, on the one hand, into the end-sac by an aperture and on the other hand, into the bladder by several apertures.

The wall of each excretory tubule is formed of a single-layered excretory epithelium whose cells have large spherical nuclei and straited border.

(iii) *Bladder*. The urinary bladder is a thin-walled sac which consists of a single layer of excretory epithelial cells. It is situated on the inner side of the end-sac. Internally, its inner wall is continued as a short, excretory a duct or *ureter*, which opens to the outside by a small renal or excretory aperture located on a papilla on the inner side of the coxal podomere of the antenna and in front of the labrum.

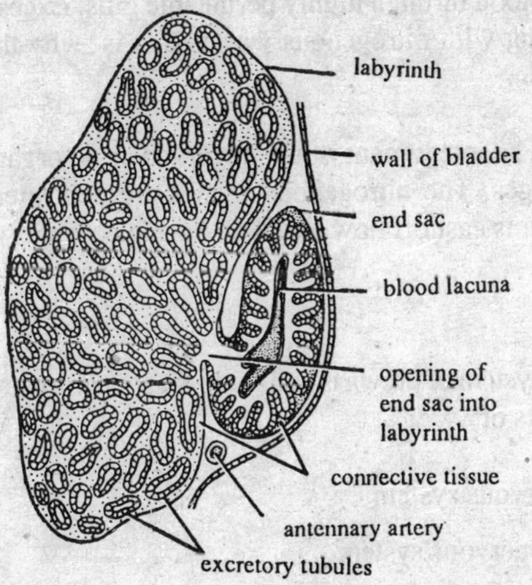

Fig. 2.32 *Palaemon* Section of Antennary gland

From the bladder, a narrow *lateral duct* runs posteriorly along the oesophagus. The lateral ducts of both the sides are joined by a transverse connective. These lateral ducts open into an elongated *renal or nephro - peritoneal sac.*

2. *Renal sac or nephro peritoneal sac.* It is a large median sac lying just beneath the dorsal shield. It covers the entire cardiac stomach and reaches the gonads posteriorly. Anteriosly it communicates with bladder of each side by the lateral duct. The wall of renal sac is made up of a single layer of flatenned excretory epithelial cells.

Physiology of excretion. The antennary glands perform two functions, excretion and osmoregulation. These glands are richly supplied with blood so that by ultrafilteration, the water and dissolved substances are passed into the end sac. The filterate passes into the labyrinths. Where selective reabsorption takes place. The useful substances are pass back into the blood. The remaining fluid, better to say the *urine,* flows into the bladders and the renal apertures. The urine contains mainly ammonium compounds, some urea and amino acids.

Osmoregulation. Palaemon is a fresh water crustacean so that due to higher salt concentration of body the surrounding water continually diffuses into the blood through highly permeable gills. Excessive water is eleminated along with nitrogenous wastes that is why the urine of prawn is hypotonic.

Integument. The integument is also regarded as an organ of excretion in young stages. The nitrogenous wastes are deposited over the integument, which is cast off now and then during ecdysis.

Nervous System

The nervous system of prawn is similar to that of earthworm in many respects. It consists of:

1. Central nervous system
2. Peripheral nervous system
3. Autonomic nervous system.

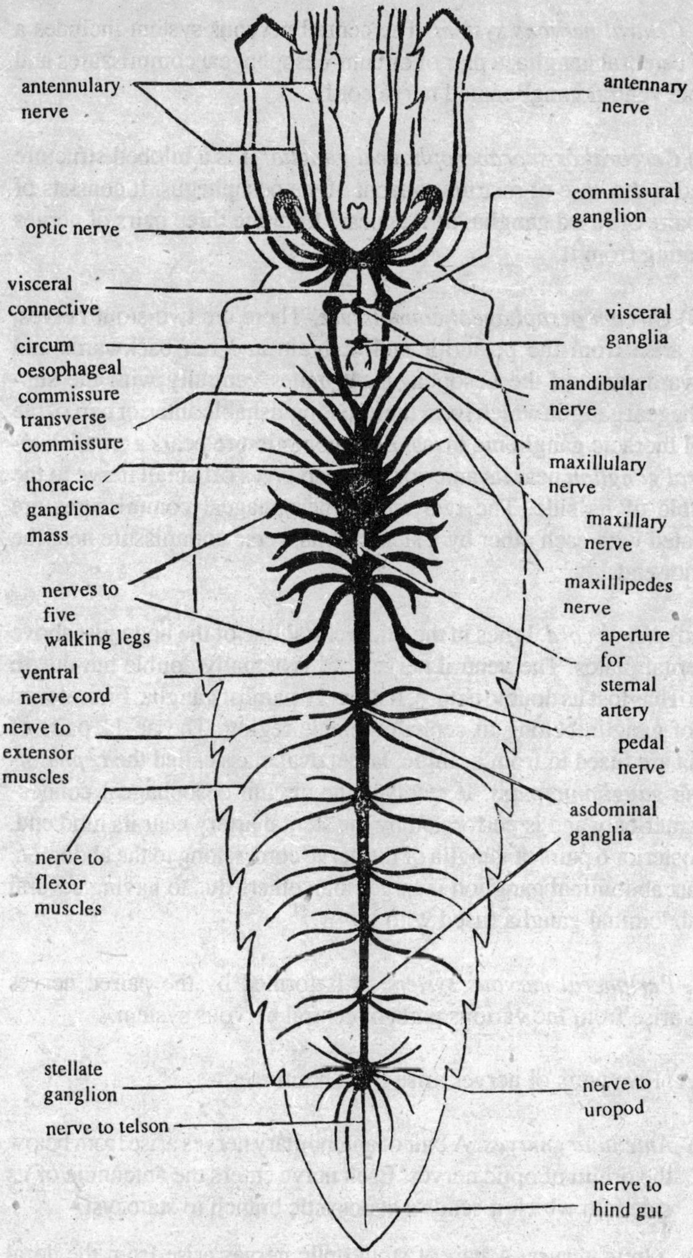

Fig. 2.33 *Palaemon* Nervous System

1. *Central nervous system.* The central nervous system includes a pair of cerebral ganglia, a pair of circum-oesophageal commissures and a double ventral ganglionated nerve cord.

(i) *Cerebral or supraoesophageal ganglia.* It is a bilobed structure situated at the base of rostrum in front of the oesophegus. It consists of three pairs of fused ganglia, as it appears from the three pairs of nerves originating from it.

(ii) *Circum oesophageal commissure.* There are two stout nerves, which arise from the posterior end of brain and run backwards and downwards around the oesophagus. It unites ventrally with the sub-oesophageal ganglia which form the indistinguishable anterior part of the ventral thoracic ganglionic mass. Each commissure bears a small *commissural ganglion* near its anterior end, and gives off small nerve to the mandible of its side. The two circum-oesophageal commissures are connected with each other by a slender transverse commissure near the posterior end.

(iii) *Nerve cord.* It lies in the mid-ventral line of the body just above the sternal plates. The ventral nerve cord is actually double but due to fusion it has lost its double nature. It bears 17 pairs of ganglia. First eleven pairs of ganglia belong to cephalothoracic region. Thesae 11 pairs of ganglia are fused to from a single, large, oval mass called the *cephalothoracic ganglionic mass.* It receives the circum-oesophageal connectives anteriosly and is perforated by the sternal artery near its hind end. The posterior 6 pairs of ganglia of the nerve cord belong to the abdomen. The last abdominal ganglion is larger than others due to having several post-abdominal ganglia fused with it.

2. *Peripheral nervous system.* It is formed by the paired nerves which arise from the various parts of central nervous system.

(A) Five pairs of nerves arise from brain are:

(i) *Antenulary nerves:* A Pair of antennulary nerves arise from below the origin of optic nerves. Each nerve enters the antennule of its side into which it sends a statocystic branch to statocyst.

(ii) *Optic nerves.* A pair of stout optic nerves arise from the doral surface of the brain, one on each side, and supply the eye-stalks.

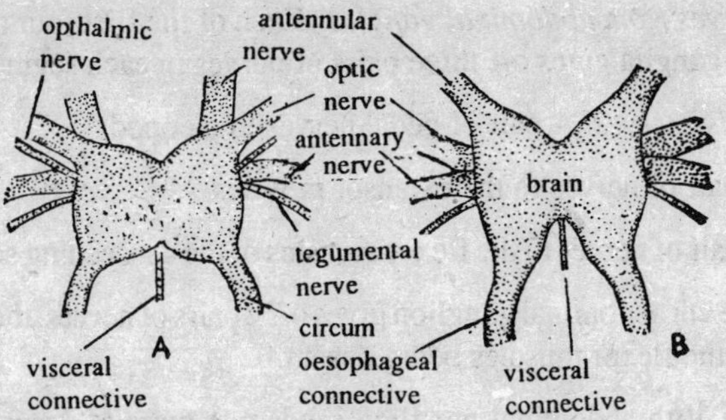

Fig. 2.34 *Palaemon* Brain A= Dorsal view B= Ventral view

(iii) *Opthalmic nerves:* A pair of opthalmic nerves arise from the brain, one on each side, and supply the ocular muscles in the eye-stalks.

(iv) Antennary nerves. A pair of stout antennary nerves originate from the ventral surface of the brain. Each nerve divides into two branches, the outer innervating the squama and the inner innervating the feeler of antenna.

(v) Tegumental nerves. A pair of slender tegumental nerves arise just behind the origin of antennary nerves. These nerves innervate the labrum.

(B) *Mandibular nerve.* It arises from the outer side of each circum-oesophageal connective and supplies the mandible of its side.

(C) *Nerves from cephalothoracic ganglionic mass.* 11 pairs of nerves are given out from the mass.

(a) *Mandibular nerves* for mandibles.

(b) *Maxillulary nerves* for maxillulae.

(c) *Maxillary nerves* for maxillae.

(d) *Maxillipede nerves.* There are three pairs of maxillipede nerves which supply to maxillipedes.

(c) *Peraeopod nerves.* These are 5 pairs in number and supplying to the walking legs.

(d) *Nerves from abdominal ganglia.* Each of the first four abdominal ganglia gives off three pairs of nerves in each segment:

(a) *Pedal nerves.* A pair of pedal nerves to pleopods

(b) A pair of nerves to the extensor muscles.

(c) A pair of nerves to the flexor muscles of the succeeding segment.

(d) The vth abdominal ganglion gives off 2 pairs of nerves, the nerves for the flexor muscles being absent.

(E) The Vith abdominal ganglion sends two pairs of nerves to the flexor muscles of sixth segment, two pairs of nerves to the uropods and two pairs of nerves to telson.

3. *Antonomic nervous system.* The autonomic nervous system comprises a few small ganglia and nerves. A small nerve arising from the mid-posterior part of the brain, bears two *visceral oesophageal ganglion,* lying one behind the other. The first ganglion is jointed with the two *commissural ganglia* by a pair of connectives. The 2nd ganglion gives off two pairs of nerves to the walls of the oesophagus and cardiac-stomach.

Endocrine System

The endocrine system comprises a group of neurosecretory cells in each eye-stalk. These cells secrete several hormones into the *sinus gland* lying adjacent to them. From sinus gland, hormone is poured in the blood. These hormones are believed to regulate - colouration of body and compou eyes; deposition of lime salts in the exoskeleton and moulting. Moulting is controled by two hormons moulting inhibiting hormone is secreted by X-organs in the eyestalks and moulting accelerating hormone by the Y - organ present beneath the adductor muscle of mandible.

Sense organs

The sense orrgans of *Palaemon* are:

1. Statocyst.
2. Olfactory setae
3. Tectile setae

4. Compound eyes.

1. *Statocysts.* The statocysts lie attached to the dorsal wall of the precoxa of each antennule and open through apertures on the dorsal concave surface. Each aperture is covered by a fold of skin. A statocyst is a subspherical cuticular sac about 1-1.5 mm wide. Each statocyst is supplied by the statocystic branch of antennulary nerve.

The cavity of the capsule possesses a heap of sand particles lined by a number of delicate elongated statocystic setae. Each seta consists of a swollen *base* and a long *shaft* which is half bent upon itself and covered with fine bristles. The base movably articulates with the wall of the sac by means of an arthrodial membrane and receives a fine branch from the statocystic nerve.

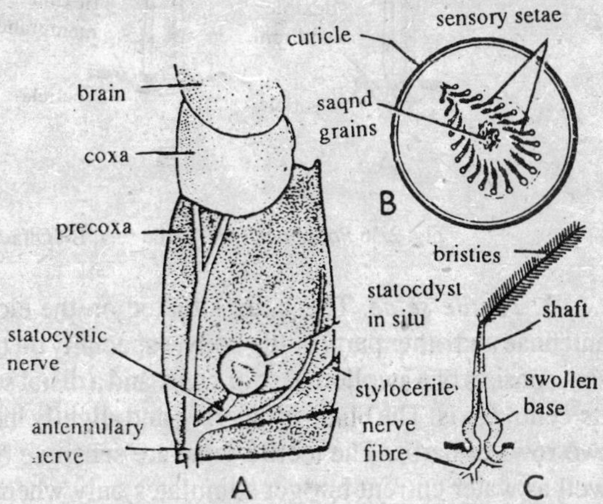

Fig. 2.35 *Palaemon* A= Statocyst in sity, B= T.S.. C= a single receptor

The statocysts are the organs of orientation and equilibrium. In the normal swimming position, the sand grains lie on the floor of the statocyst under the influence of the force of gravity. When the body of a swimming prawn gets tilted to one side it causes the displacement of sand particles in the same direction. This presses the sensory setae and stimulates them. The stimulated setae transmit this feeling to the brain and the prawn knows about its position and balances itself accordingly.

2. *Olfactory setae.* These are also known as aesthetases, occur on the middle small feeler of each anternule. Each setae consist of a basal segment or *shaft* and is attached segment or *blade*. Shaft is hollow and

is attached to the integument while the later (blade) is free and bluntly rounded and is covered with a thin membrane. Small nerve fibres from the antennulary nerve supply these setae.

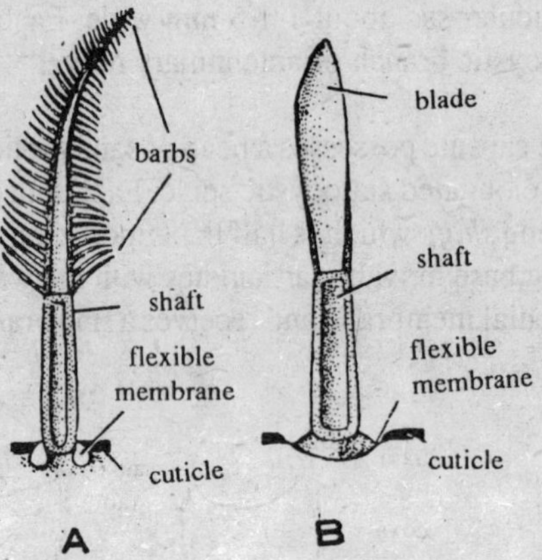

Fig. 2.36 *Palaemon* A=Textile seta, B= Olfactory seta

3. *Tactile setae*. These are situated on the elongated feeler of the antennae and other parts of the body, especially on the appendages. Each seta consists of a swollen base or *shaft* and a distal segment or *blade* with its central axis. The blade is tapering and slightly inclind distally and has two rows of barbs. The tectile setae are sensitive both to the surface as well as water current but get stimulates only when moved.

4. *Compound eyes*. There is a pair of compound eyes, which are black and hemi-spherical. Each eye is situated at the base of short,

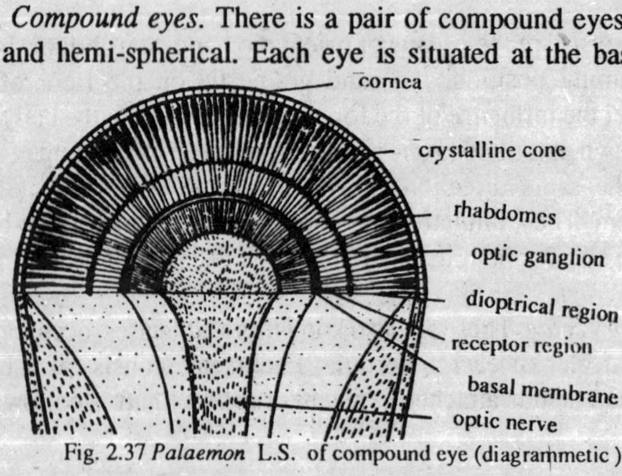

Fig. 2.37 *Palaemon* L.S. of compound eye (diagrammetic)

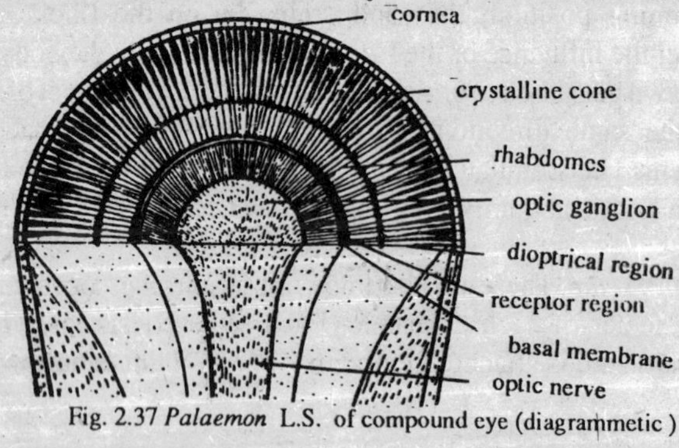

movable and two jointed stalk. Each eye is contained in an orbital notch of the rostrum. Each eye is actually a composite structure being composed of a large number of visual units called as *ommatidia*. There are about 2500 such ommatidia in a compound eye. The surface of the eye is roughly hemi-spherical and is divided into a large number of small squase facets . Each facet corresponds to a single ommatidium. All the ommatidia are similar, displayed radially, lying side by side and separated by a dark pigment cells. Each ommatidium consists of following parts:

(i) *Cornea.* The eye is protected externally by a transparent covering of cuticle known as cornea. It is distinguished into facets like those of graph paper.

(ii) *Corneagen cells.* These arc flat one pair and of epidermal in origin. They secrete a new corneal layer after ecdysis.

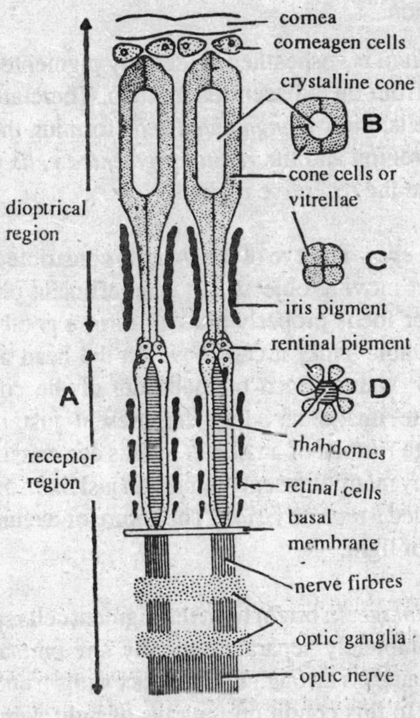

Fig. 2.38 *Palaemon* (See figure for Caption)

(iii) *Cone cells.* Beneath the corneagen cells lie four elongated cone cells or *vitrellae,* which constitute a transparent, homogenous *crystelline cone.* The inner ends of cone cells are long and tapering.

This part of the eye from the cornea to the posterior end of the cone cells is called *dioptrical region* as it is concerned with the focusing of the light on the under lying region known as *receptor region.*

(iv) *Rhabdome.* The inner ends of cone cells lie upon an elongated, spidle-shaped, transversely striated structure known as rhabdome. It is secreted and surrounded by a group of seven elongated *retinal cells.* The inner end of the rhabdome are supported by a basal membrane beyond which they are continuous with the nerve fibres. The rhabdome and retinal cells from the *receptor region.*

(v) *Nerve fibres.* The ommatidia are innervated by nerve fibres of the optic ganglion.

Each ommatidium is ensheathed by densely pigmented cells which optically separate from the adjacent ommatidia. There are two sets of these pigmented cells: the *iris pigmented cells* forming the *iris sheath.* round the dioptric region and the *retinal pigmented cells* that form the *retinal sheath* round the receptive region.

Working of the eyes. The eye of prawn is so constricted that while it can swiftly spot out moving objects and is an efficient photoreceptive organ, it can neither focus properly nor can form a good clear image. Mounted on a movable stalk, it can move on the head and gives the animal almost 360^0 vision. Each ommatidium of the compound eye produces a separate image of an object present just infront of it. Therefore, the image formed of an object by this eye consists of several pieces contributed by many adjacent ommatidia just like a mosaic art net. That is why it is called a *mosiac vision.* The nature of the image depends upon the intensity of light.

(a) *Apposition image.* In bright light the pigment cells spread in such a way that they completely separate optically one ommatidium from adjacent. No light can pass through the chamber walls from one ommatidium to the other. In this condition the rays of light, which strike the cornea obliquely are absorbed by the pigment cells without producing a

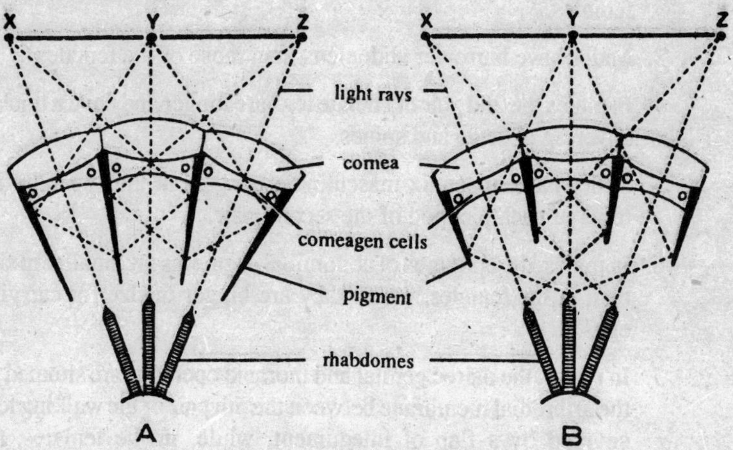

Fig. 2.39 *Palaemon* A=Apposition or mosaic vision, B= Super-position image.

visual effect. Only those rays of light, which pass directly through the centre of cornea, can traval through the ommatidium and reach the rhabdome to from an image of a part of object. As a result the complete image formed is a mosaic of several components places in juxtaposition. Such as image is called *apposition image*.

(b) *Super Position Image*. In the weak light, the pigment cells migrate partly towards the outer and partly towards the basal ends of the ommatidia. When this is the condition, even the oblique rays of the source of light are capable of forming a point of image, after passing through a number of ommatidia. Thus, a combined image is thrown on the retinular layer.In this way, an erect superposition image is formed as the rays refracted by several crystalline cones are superposed on the focus of retina. The superposition image is less distinct in detail than the apposition image, but it is more sensitive to weak intensities of light.

Reproductive System

The sexes are separate and there is a well marked sexual dimorphism. The males and females differ in several respects.

1. The males are larger than females.

2. Males have thoracic legs more closely arranged than in the females.
3. Males have narrower abdomens then those of the females.
4. In males the 2nd pair of chelate legs are longer and have a thicker covering of setae and spines.
5. Males have appendix masculina arising from the appendix interna of each pleopod of the second pair.
6. In males, the epimeres of abdominal segments are smaller in size than in the females, where they are bigger in size for carrying eggs.
7. In males, the paired genital and thoracic openings are situated on the arthrodial membrane between the 5th pair of the walking legs covered by a flap of integument, while, in the females, the generative apertures are situated on the coxae of 3rd pair of the walking legs.

Male reproductive organs. The male reproductive system comprises:

1. Testes.
2. Vasa deferentia
3. Vesicula seminalis.

1. *Testes.* The testes are median dorsal structure situated in the posterior part of the thorax above the hepatopancreas and below the pericardium. They extend anteriorly upto the renal sac and posteriorly upto the first abdominal segment. These are soft white structures which are fused anteriorly to from a single lobe, while at the posterior end these remain in close contact. The two testes enclose a central space or gap which gives way to the *cardiopyloric strand* connecting the heart to the cardiac stomach.

Each testis consists of a large number of coiled, hollow thin walled *seminiferous tubules* buried in connective tissue. The cavity of each tubule is lines by a single layer of *epithelium,* the cells of which undergo *spermatogenesis* to form the *spermatozoa*. A mature sperm consists of a

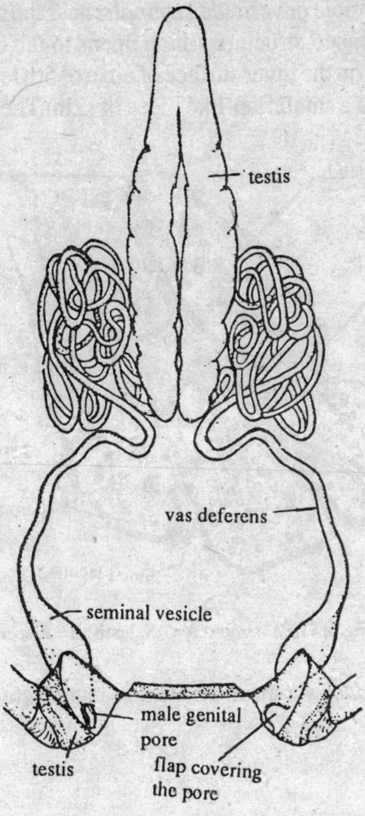

Fig. 2.40 *Palaemon* Male reproductive system.

rounded cytoplasmic body, containing a large, dark crescentic nucleus, and a tail-like blunt process.

2. *Vasa deferentia.* From the outer surface of the posterior end of each testis arises a very long and narrow tube, the *vas deferens*. It is differentiated into two parts: the proximal highly convoluted part which emerges out of the testis and the distal straight part which runs vertically downwards between the abdominal flexor muscles and the thoracic wall. Each runs inwards to open into the vasicula seminalis.

3. *Seminal vesicles.* The distal part of each vas deferens enlarges to form a seminal vesicle or *vesicula seminalis* near the coxa of 5th walking leg. It is a club shaped structure which opens to the outside through the male genital pore on the inner surface of coxa of 5th leg. The male genital pore is covered bu a small flap-like piece of skin. The vesicula seminalis stores the sperms.

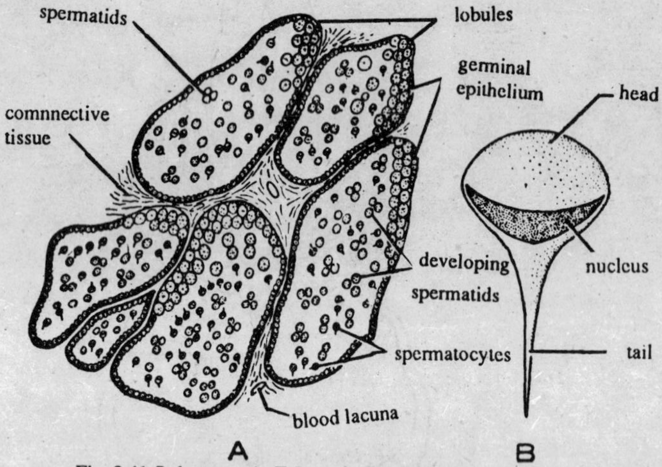

Fig. 2.41 *Palaemon* A=T.S. testis, B= A sperm

Female reproductive system. The female reproductive system comprises following organs:

1. Ovaries
2. Oviducts.

1. *Ovaries.* A pair of ovaries which are compact and sickle-shaped bodies touching each other at both the ends but leaving a gap in the middle for the passage of the cardiopyloric strand. The shape and size of the ovaries vary with age and the season of the year. Each ovary is enclosed within a membranous capsule and is made of numerous radial rows of *ova* in various stages of development. The immature ova lie towards the centre while the mature ova towards the surface of the ovary. The latter are large nucleated cells with planety of yolk material.

2. *Oviducts.* From the middle of the outer surface of each ovary arises a short thin-walled tube known as *oviduct.* Its anterior end is broad and forms the *oviducal funnel.* It lies opposed to the surface of ovary. The

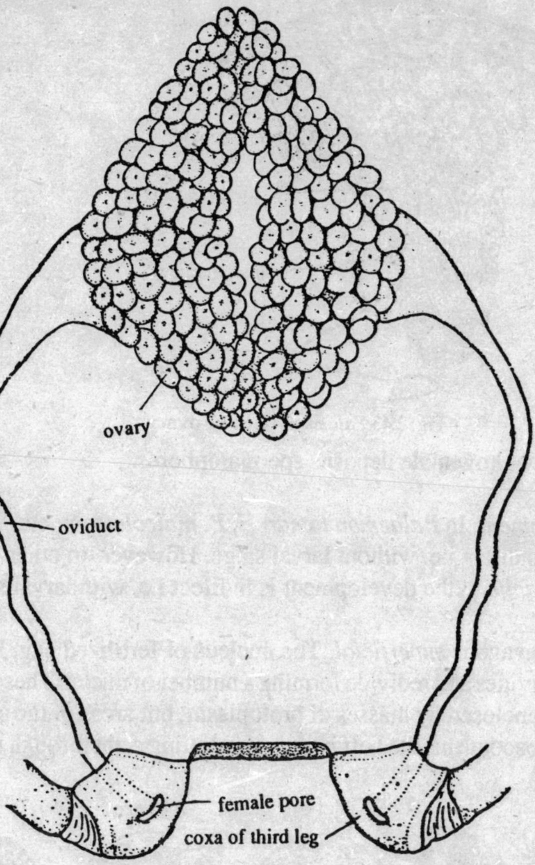

Fig. 2.42 *Palaemon* Female reproductive system

oviduct runs vertically downwards to open to the exterior by *female genital pore*. It is situated on the inner side of coxa of 3rd walking leg.

Fertilization. Breeding season in *Palaemonis* from May to July. During the season, the females carry eggs attached to their pleopods. Except in breeding season, the females are found without eggs. Fertilization is external. When ova are laid, they are attached to the setae on the pleopods with the help of a sticky secretiopn of glands situated both on the appendages and on the body segments. The male, later on, deposits sperms near the genital openings, where fertilization takes place. It is not

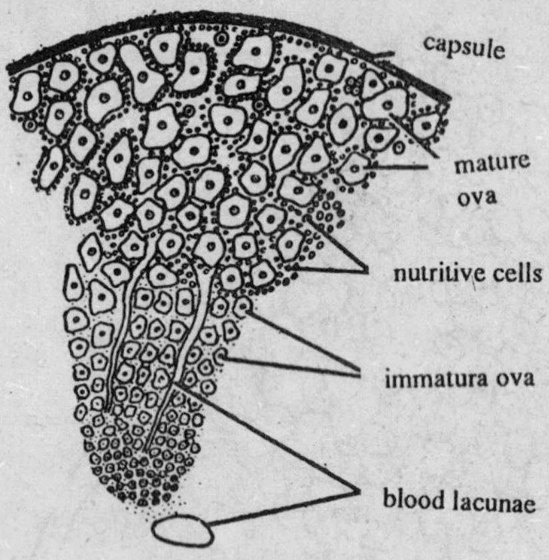

Fig. 2.43 *Palaemon* T.S. of ovary

yet known that how male deposits spermatophores.

Development. In *Palaemon lamarrei, P. malcolmsonii* etc. The development is direct i.e. without larval stage. However, in other prawns (*Penaeus, Lacifer*) the development is indirect i.e. with larval stages.

The cleavage is *superficial*. The nucleus of fertilized egg, lying in the centre, divides and redivide forming a number of nuclei. These nuclei are later on enclosed by masses of protoplasm, but areas of the separate cells do not become marked off by furrows cutting right through the egg.

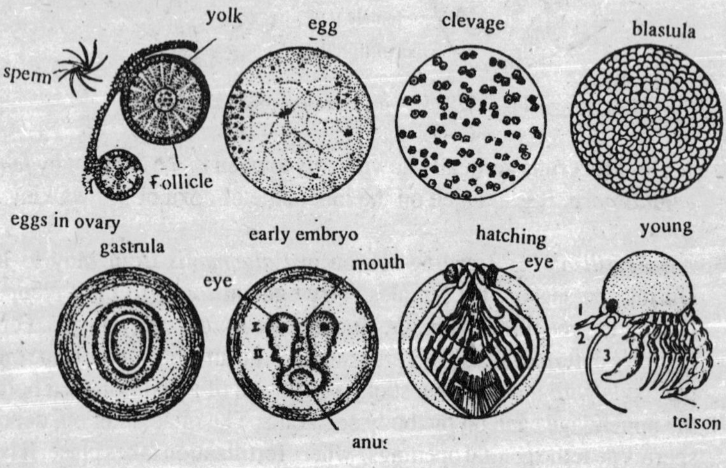

Fig. 2.44 *Palaemon* Showing development

As a result of segmentation, a blastoderm is formed, which appears at first on the ventral side of the egg. Its formation begins at one point on the surface and proceeds gradually from this point, which always represents the futute ventral side of the egg. Gastrulation starts fron the posterior end of the ventral side which is marked by the first formation of the blastoderm.

Gastrulation. A small invagination develops at a time when the blastopore is not completely formed. This invagination marks the first step towards gastrulation and small *archenteron* is formed with its opening, the *blastopore*. When the blastopore closes, the cells of the endodermal vesicles lose their epithelial nature. From the lateral walls of the endoderm vesicle arise elements which form the mesoderm at a later stage Endoderm is formed by the cells, which arise from the floor of endodermal vesicle and pass on to the yolk and multiply therein. Each of these endodermal cells swallows the surrounding food yolk, with the result the yolk mass breaks into smaller spheres. Afterwards the midgut is formed by the inter-penetration of the food yolk by the cells of gastrula. Fore gut and hind gut also arise by ectodermal invagination forming stomodaccum and proctodaeum. Both reach the midgut and finally fuse.

When the mesoderm appears for the first time, it sonsists of a mass of cells near the point of origin. These cells multiply rapidly and spread out apparently in an irregular manner between the ectoderm and the food yolk.

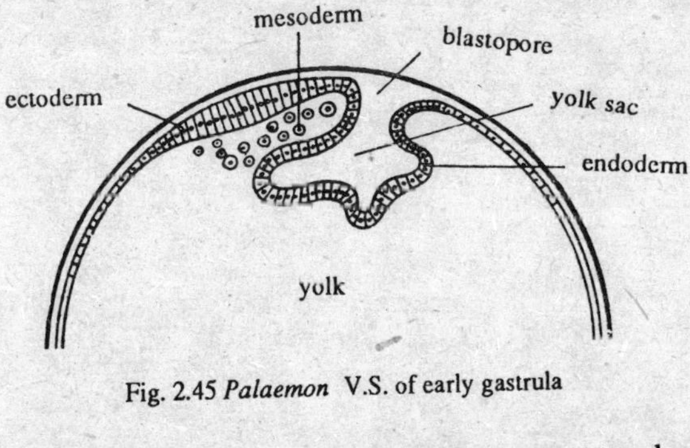

Fig. 2.45 *Palaemon* V.S. of early gastrula

The ectoderm gives rise to the external integument and an internal skeleton. Nervous system develops from ectodermal thickening. Vascular system is developed from the mesenchyme. Growth of the main parts of the embryo in size, during which the entire yolk is consumed, results in a small individual which is similar to adult except in size. After several moults the adult is formed.

Power of Regeneration

Prawns and certain other crustacians like lobsters, crabs, cryfishes etc., have a power of regeneration which is greater in young stages but very slow in adults for example, if the tip of eye is removed, normal regeneration takes place. If the entire eye stalk along with optic nerve is removed, an antenna like structure is formed. Such type of regeneration in which normal organ is not developed, is called *heteromorphosis*.

Crustaceans have the power of autotomy. If a thoracic leg is broken in some mishap, its distal segments are spontaneously thrown off to feed the enemy.

3

PRAWN FISHERY

Prawn, a crustacean, commercially known as shrimps, is considered by most people a fish. Due to its nice taste, it is supposed to be a cherished delicacy to serve as food throughout the world. Prawn fishery implies capture of prawns from their natural resources and also their culture. It has sufficiently advanced in in India. There are different varieties a prawn inhibiting sea, estuaries and fresh waters. Production of prawn in India is second to U.S.A. and it is also important because it is an important item of our export and the annual foreign exchange returns from 70-80 crores of rupees.

Types of Prawn. As mentioned earlier that prawn inhibits all sorts of water i.e. marine, esturies and fresh water of temperate and tropical countaries. Prawn fishing in marine water is mainly restricted to wes coast. Kerala and Bombay coasts are the chief centres for the production of marine prawns. Among marine prawn, the older specimens line upto 30 metres deep and the younger one up to 18 meters deep. The important species of prawns found in coastal region and *Penaeus monodon* (the giant tiger prawn), *P. indicus* (Indian prawn), *P. mergniensis* (banana prawn), *Metapenaeus dobsoni* (yellow prawn), *M.affinis, M. monoceros* etc. are very common. Among thesc P.indicus occurs most commonly and constitute the bulk of production. It measures upto 20 cm in length. The largest being *P. monodon*, measuring 30 cm in length but it is not very common. *Metapenaeus dobsoni*, *M. affinis* and *M. monoceros* are most suitable species to be cultured in India, Pakistan, Philippines, Malaysia, Indonesia etc.

The species of estuaries prawns are the same which occur in marine water except *Penaeus stylifera*. Which is exclusively marine. The important areas of prawn production in estuaries are sothern half of Kerala Coast. Several lakes like Chilka, Colliar, Ennore, Pulicat etc. are also some areas of prawn production.

The exclusive freshwater species is *Macrobrachium sabriculs*, where as *Metapenaeus monoceros* and *Penaeus monodon* are feshwater as well as marine, for freshwater culture, giant fresh water prwans.

Macrobrachium resenberghi, *M. malcolmsoni* are best suited.

Fishing. The method employed for fishing prawns are somewhat similar to that of fishes. Fresh water prawn are caught round the year and so also the marine prawn but there are certain peak seasons. Along west coast, it is November to May, offshore fishing is most profitable from July to October. Along east coast from December to August, various types of nests are used for prawn fishing, amongst which the commonest type used is boat siene. In estuaries, a peculiar type of conical net, supported by a pair of stakes is being used. Other types of nests like wall net, drag net etc. are also used for this purpose.

Prawn culture or farming. These are two methods for prawn culture: (i) The old and conservative process. Commonly called as *traditional method* is widely used in India. The ponds situated near the estuarine and brackish water are selected for the purpose. The ponds remain connected with the outer main tidal stream by sluice gates. These are kept open during high tide when larvae get in. During low tide they are open but the larvae are prevented from escaping. After growing upto a marketable size, the field water is filtered and the prawns are collected. During their growth the larvae are provided with natural diet. However, in Philippines, instead of a natural diet, artificial diet is provided. Another method of collecting is to set up bag nets or bamboo traps near the sluice gates. During night electric bulbs which attract shrimps are also used for quick trapping.

(ii) Philippines, Japan, U.S.A. etc. are employing modern technique for culturing. This method is called *intensive prawn culture*. It includes the rearing of post larval stages of prawn and its breeding and spawning at desired places by providing the artificial environment. Matured males and females are kept in small concrete tanks where temperature and oxygen concentration are maintained artifically. After spawning the adults are removed. The early larval stages are fed on phytoplanktons like diatomes where as the post-larval stages are provided brine shrimp larvae, worms etc. as food. On attaining the desirable size, the larvae are transferred to the production tanks. These tanks have two openings which are connected to the main stream of water through water pipes with the help of hydraulic pumps, the water in poud is made to circulate. This method gives much more production than traditional method.

In India intensive prawn culture is in experimental stage. Experiments performed by Kerala state at Narakkal, Vypcan Island have shown an annual production of 363 kilogramms per acre.

Preservation and Processing. Since the prawn is an important article for export, their preservation and processing is necessary. If distance is short, they are packed between laryers of ice either entirely or after removing cephalothorax. For long time preservation two methods are employed:

(i) Total drying

(ii) Semi drying

(i) In total drying, the entire prawn are dryed in tube sun for several days. In some places, they are boiled are thrased to remove the shell. Now they are packed for marketing.

(ii) In semidrying techniques the prawns are boiled in 6% brine for just two minutes. After that the shell is removed and the prawns are dipped in saturated salt solution for half an hour. The process is follwed by drying in the sun, but the drying is stopped before the flesh gets too hard. Such semidried prawns when socked in water give the original taste after months.

Besides other methods like chilling and freezing, pulp making, pickling, smoking and canning are also employed.

Prawn manure. It includes semi-dried prawns and heads, tails and chitinious shells of prawns. It contains 5-6% nitrogen and phosphates and chitinous body shells if dried and powdered, provide prawn meal a protein rich poultry feed.

Pollution. Prawn fishery is affected by pesticides, effluents, metals and oils.

4

LARVAL FORMS IN CRUSTACEA

The embryonic development is followed by metamorphosis which may be either very well marked or supressed. In most of the crustaceans the egg hatches to produce Nauplius larva which in free swimming but in some cases the nauplius stage is passed within the egg so that there is no free swimming nauplius larva produced after hatching. It crabs, fresh water prawns and cryfishes the development is direct i.e. without larval stage. In others several larval forms are met within crustacea and special terms are applied to each one of them. Following larval forms are met within crustacea:

Nauplius Larva

The nauplius larva is oval in shape behind any sign of external segmentation. It is characterised by the presence of 3-pairs of appendages. The first pair later on becomes antennules and the 2nd and 3rd pairs becomes antennae and mandibles. The antennules are uniramous where as the antennae and mandibles are biramous. The antennae are provided with jaws whereas the mandibles are without jaws. Thus the antennae have masticatory function. The median eye is the only organ of vision. Besides eye there is a pair of antennal glands. The mouth is guarded by a large labrum and the alimentary canal is straight and made up of foregut, midgut and hindgut. The anus may or may not present at the time of hatching. A heart is still not formed. At the posterior end there are two pairs of setae. The nauplius swimms in water by the stock of all the appendages. In Branchiopoda and Ostracoda the nauplius larva undergoes a series of moulting or ecdysis to becomes the adult. During gradual process the adult characters are gradually attained.

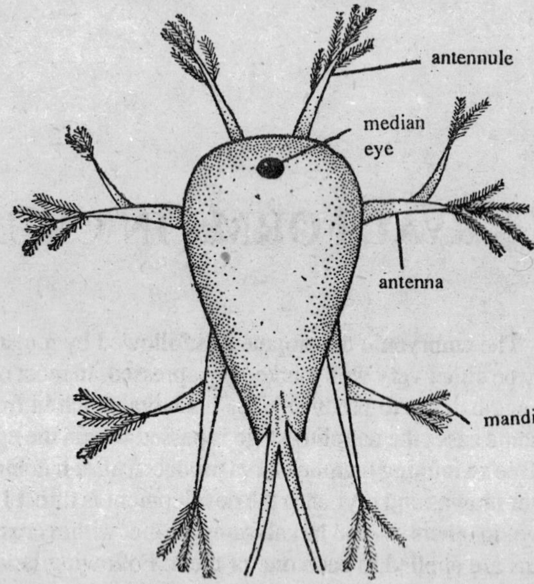

Fig. 4.1 Nauplius larva

In order Cladocera the development is direct so that the nauplius stage is passed in the egg and when the egg hatches an adult type of crustacean is produced.

In subclass Ostracoda a bivalved nauplius is produced without biramous appendages i.e. all the appendages are univamous.

In subclass Copepoda immediately after hatching a nauplius is produced and after moult it is convested into metanauplius larva.

Metanauplius Larva

It is characterised by the presence of several pairs of appendages. Besides 3-pairs of each antennules, antennae and mandibles, it develops another four pairs of appendages. Which become the maxillulae, maxillae and first two pairs of maxillipedes of the adult. Each mandible acquires a masticatory process.

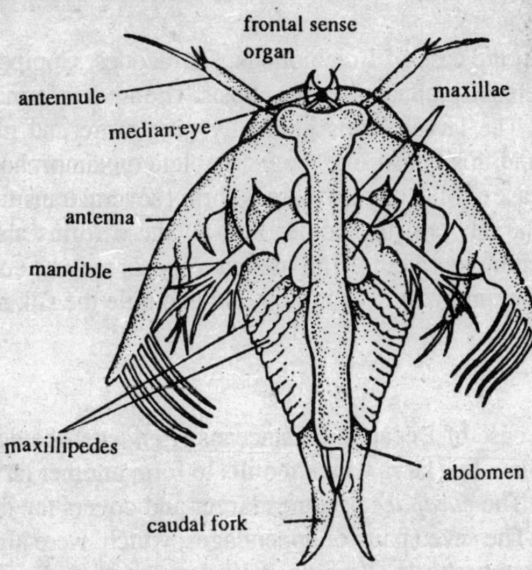

Fig. 4.2. Metanauplius larva

The body consists of an oval cephalothorax, an elongated trunk-region and abdomen, which terminating in a caudal fork provided with setae.

Cypris Larva

In subclass Cirrpedia, the free swimming stage is nauplius. The larva undergoes a moult which gives rise the cypris larva. Tdhe cypris larva is characterised by the presence of bivalved shall like that present in Ostracoda. In cypris larva the bases of antennule are modified into attachment disc which gives attachment to the larva added by the secretion of cement glands. There is a pair of compound eyes near the median eye to detained from nauplius larva. Antennae and mandibles are absent, sometimes the mandibles are rudimentary. 6-pairs of thoracic appendages are present in thoracic region. The abdomen is very short and ends into a pair of caudal fork or style. After swimming for some time, it attaches itself to some object with the help of attachment discs and secretion of cement glands and get transformed into pupa. Later the pupa becomes adult. In the parasitic crustacean like *Sacculina* free swimming nauplius and cypris are produced but the cypris larva attach itself to a crab and gradually becomes a tumover like adult.

Thus in the case of Brachiopods, Ostracodes, Copipodes and Cirripedes, there is a gradual metamorphosis. The adult is formed by a series of moults. The larval forms gradually give more and more of adult character and this is just like the incomplete metamorphosis of insect. But in the case of Malcostraca there is abrupt several transitional changes from one larval form to another and all these forms are very much different from their adults. The metamorphosis may be compared with complete metamorphosis of insects, for example the silk moth.

Protozoea.

In the case of Decapod crustaceans (*Penaeus*) the egg hatches to produce a nauplius larva which moults to form another larval stage, the *protozoea*. The *carapace* becomes larger and covers the dorsal surface anteriorly. The seven pairs of appendages, which were already formed in the metamorphosis, become well-developed and are capable of functioning, The antenae are still locomotory organs. Antennule is 4-jointed, antenna possesses a 3-jointed endopodite and a four-joint exopodite. The mandibles are palpless, toothed and are *masticatory* in function. Abdomen is still unsegmented and without appendages. However, rudiments of the remaining posterior six thoracic segments are also formed. Later on, the third pair of thoracic limb also develops, which forms the third maxillipede of the adult. Anterior five abdominal segments are also disticnt. Formation of paired eyes has also begun.

The protozoea again moults and gives rise to the zoea larva.

Zoea larva.

The body of zoea has a distinct, highly developed caphalothorax and distinctly 6-segmented abdomen. The cephalic appendages are will developed. Abdominal appendages are lacking. Zoea is characterized by the absence of mandibular palps. Eyes are large and compound situated on movable stalks. Abdomen has forked telson. The carapace is produced into long spines. A *frontal spine* is situated between the two eyes, a *dorsal spine* and a pair of lateral spines are also present.

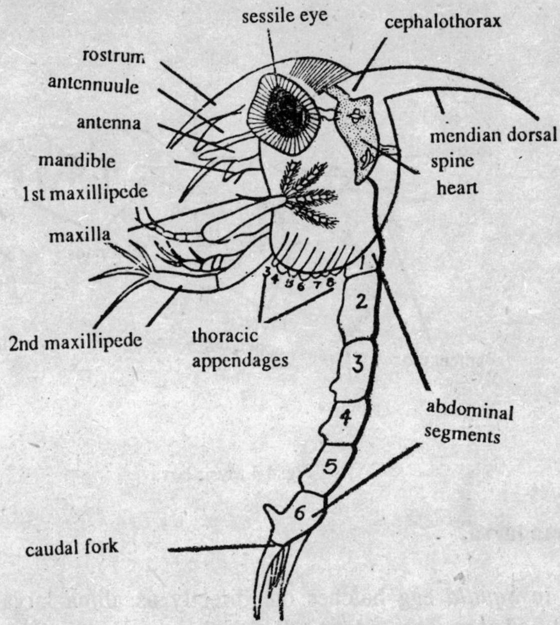

Fig. 4.3. Zoea Larva

The stage at which these free swimming larval forms begun varies vary much. In some cases the free swimming stage is nauplius but in some other cases nauplius stage is passed withing the egg, so that immediately hatching stage is metanauplius as in *lucifer*. In the case of *Sergestes* the egg produces the protozoea and in higher decapods zoea is produced immediately hatching.

Mysis or schizopode

In Peanaeus, zoaea larva instead of changing into megalopa changes into mysis which is very similar to the adult *Mysis*. It possesses five pairs of posterior thoracic legs whicha are biramous with flagellar exopodite and help in locomation. The abdomen is elongated with well developed biramous plepods. The mysis larva metamorphoses into the adult prawn.

In lobsters the egg hatches into the mysis larva.

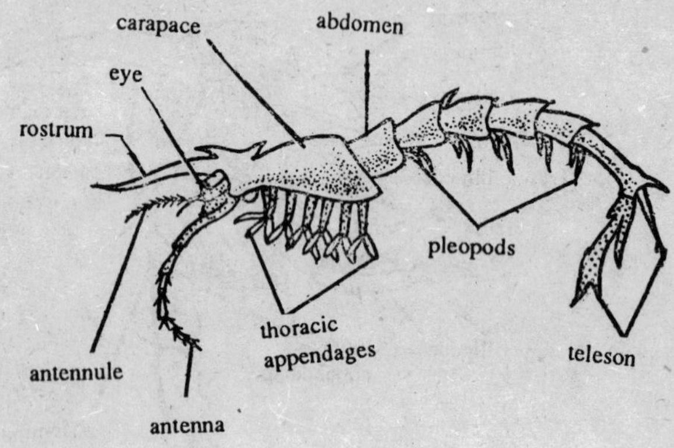

Fig. 4.4 Mysis larva

Alima larva.

In *Squilla* egg hatches out directly as alima larva, which is a modified zoea. The body is cylindrical with a short and broad carapace.

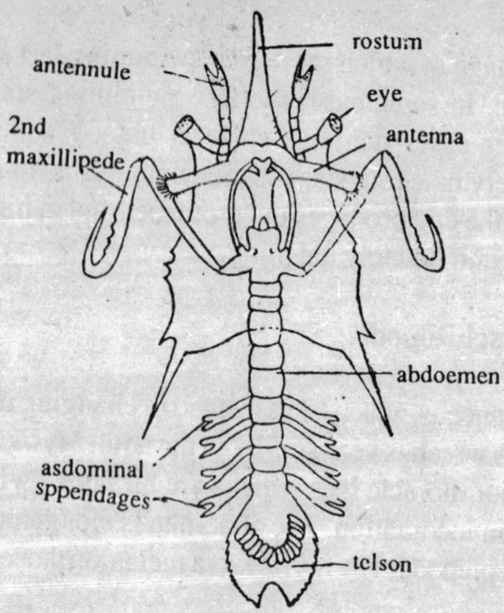

Fig. 4.5. Alima larva

The larve differs from zoea in the armature of the telson and the second maxillipede. Abdominal appendages are formed in the form of buds. The larva is pelagic. It is very transparent just like a piece of glass and is found in large numbers in plankton.

Metazoea Larva.

In the case of *Eupagurus* (hermit crab) the hatching takes place at zoea stage. The older zoea also called metazoea has well developed third maxillipede which is biramous and swimming organs. The 6-pairs of abdominal appendages also appear in the form of buds.

Calyptopsis.

In Euphausiacea, one of the larval stage is called calyptopsis. It is similar to zoea larva in all respects except that the paired eyes are sessile i.e. not stalked.

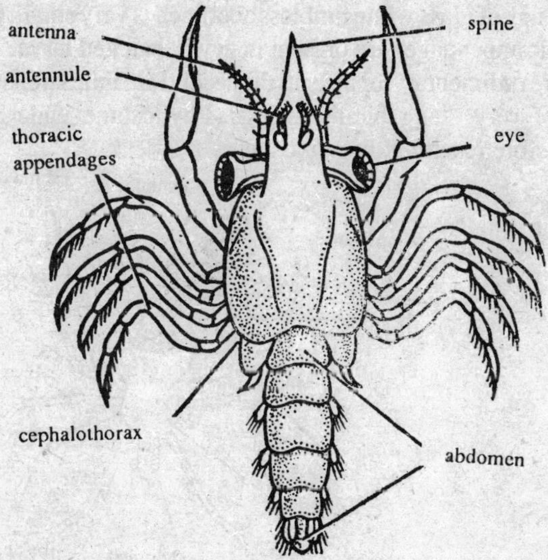

Fig. 4.6 Megalopa larva

Megalopa larva.

In some crabs, the zoea larva moults into megalopa larva. It is a crab like creature and the cephalothorax is broad but without spines. Eyes are stalked and large. The thoracic appendages are 5-pairs of walking legs. The abdomen is also well formed and bears biramous pleopods. The megalopa swims on the surface of the water leading a pelagic life but, later on, sinks down to the bottom and moults into a typical crab.

In hermit crab, the metazoea leads to glaucothoe stage. It corresponds to the megalopa stage of Branchyura. It has a large symmaterial abdomen and adult appendages.

Phyllosoma larva

In rock lobster, *Palinurus*, and its allies hatching takes place as phyllosoms larva or *glass crab* which is a greatly modified mysis state. It is strange looking, leaf-like, transparent, with eyes at the ends of long stalks. The broad depressed cephalic and thoracic shields have the transparency of glass while limbless abdomen is very small. Only 6-pairs of thoracic appendages are present in newly hatched larva. The maxillipedes aqre rudimentary or absent, the second are uniramous followed by 4-pairs of very long biramous legs. Phyllosome undergoes several moults before reaching the adult stage.

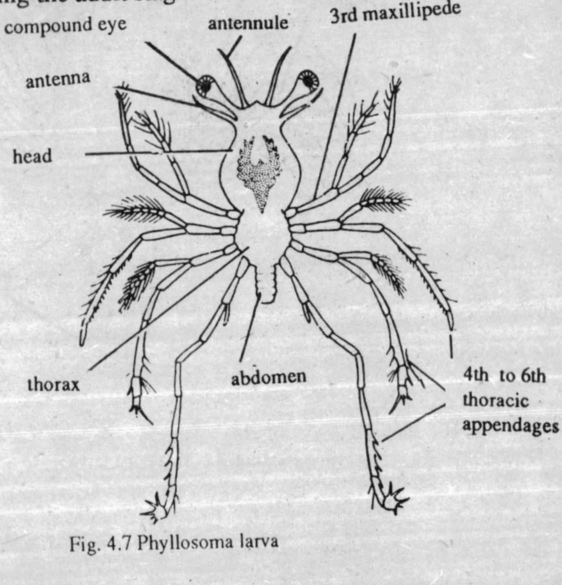

Fig. 4.7 Phyllosoma larva

Significance of larval forms.

Life history of higher crustaceans show that as the later stages are added, the earlier stages are supressed. The nauplius occurs as a free-swimming larva in lower crustaceans but in higher forms hatching takes place either mysis or zoea stage and the nauplius stage passes into egg. In certain respect the nauplius features are as follows:

1. The preoral position of antennal.
2. Presence of masticatory endite at the base of antenna.
3. Presence of biramous mandible.
4. Presence of median eye.
5. Frontal sense organs.

The biramous character, also found in some copepodes and ostracods. It appears probable that the nauplius was establishing as a larval form at a period of development before the divergence of the ancestral groups of crustacean and probably that is the reason why the nauplius stage appear in some form of life-history of some of the crustaceans. None of the earlier adult crustacean shows a nauplius like form but rather a multisegmented form of the annelidian type. Similarly none of the primitive crustacean that have none of the zoea larvae but zoea like adults. As mysis occur in the development of decapod and resembles each other inspite of the fact that their adults are entirely different. This shows the structure of crustacean ancestors. If we accept that "Ontogeny recapitulates phylogeny" i.e. the various larval forms give ancestral history during development but this is considered only partially correct. The explaination that has been given to divert these forms is that they have adapted to special environmental conditions at diffeent time, however, some phylogenetic significance attached only to nauplius larva because of primitive features of crustacean stock. The other larval forms have no phylogenetic significance and have developed in accordance with influence of environment which were different from time to time.

5

PERIPLANETA AMERICANA

The cockroach is a familiar type of the class Insecta belonging to order Blattaria. These are very ancient insects and are found practically all over the world except the polar region. There are about 2600 species of cockroaches of which the common species in India are *Periplaneta americana, Periplaneta australasiae, Blatta orientalis* and *Blatta germanica. Periplaneta americana* was originally designated as *Blatta orientalis by Linnaeus* (1958), later changed to *Blatta kakeralac* by *De Geer* (1973) and finally the present name was given by *Burmeister* (1838).

SYSTEMATIC POSITION

Phylum	-	Arthropoda
Class	-	Insecta
Subclass	-	Pterygota
Division	-	Exopterygota
Order	-	Blattaria
Family	-	Blattidae
Genus	-	*Periplaneta*
Species	-	*americana*

Habits and Habitat

They are generally found in warm damp places particularly in kitchens, cupboards, hotels, drains, ships, trains and even sewers. Cockroaches eat everthing fit for human consumption, besides they will eat leather, paper, cloath and wool, thus they are omnivorous. Sometimes when the food is not available they turn cannibal and eat each other. Cockroach is nocturnal i.e. it comes out to feed at night. Cockroaches are cursorial i.e. run very fast though they can fly also. The eggs are laid in oothecae. The young ones hatching from oothecae resemble the adult in mode of life except that they are smaller in size.

External Characters

Shpae and size. The body of cockroach is narrow elongated, bilaterally symmetrical and dorsoventrally flatenned. It is about 4 cm in length and one cm wide

Colouration. The colour is reddish brown. The tergum of the first thoracic segment bears two dark patches surrounded by a light brown margin.

Exoskeleton. The entire body is covered by a hard brown coloured chitinous exoskeleton. Each segment of the body has hardened plates called *sclerites* which are joined to each other by thin flexible *articular membranes*.

Division of body. The body is divisible into three distinct region:

1. Head

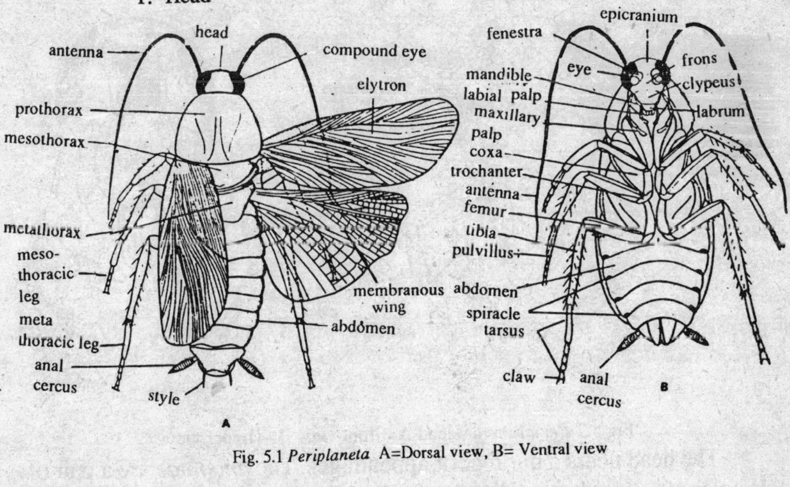

Fig. 5.1 *Periplaneta* A=Dorsal view, B= Ventral view

2. Thorax
3. Abdomen.

1. *Head.* The head comprises the anterior body region of cockroach. It is sclerotized capsule formed by the fusion of several sclerites, often distinctly seen in the embryo. It lies at right angle to the longitudinal axis of body and is highly movable in all direction due to a flexible neck.

The head bears large compound eyes which are broad dorsally. A dark sclerite covers the top of the head and extends downwards between the eyes. It is called *vertex*. In the nymph the vertex is divided into two *epicraminal plates* by an inverted Y shaped *epicranial suture*. During moulting of the numph, the head capsule splits at the epicranial suture, and the epicranial suture disappears in adult. The unpaired triangular plate, lying between the arms of the epicranial suture is the *frons*. Below the frons is *clypeus* and *labrum*. Covering the sides of the head and lying below the compound eyes are cheek sclerites or *genae*. In the angle between the eyes and antennae are two small pale coloured areas called *fenestrae*, which are undeveloped *ocelli*. At the back of the head is a large rectangular *occipital formen* bordered by an arched sclerite, the *occiput*.

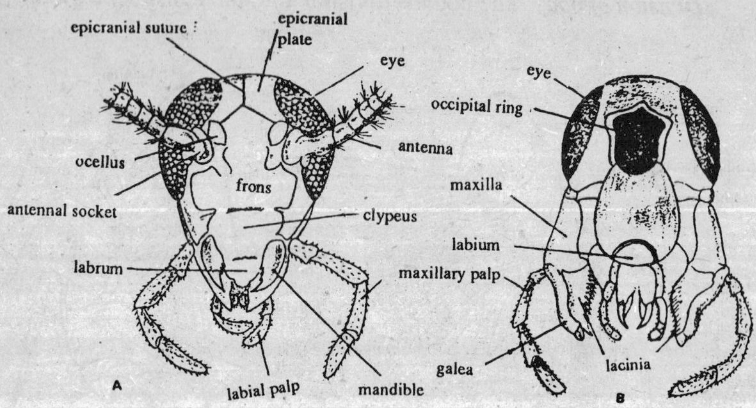

Fig 5.2 *Periplaneta* Head A=Front view B= Hinder view

The head bears a number of appendages. The *antennae* are a pair of long, multijointed, thread-like, movable structures attached infront of the eyes. Each antenna articulates in membranous pit, the *antennal socket* and has three parts a basal large segment, the *scape*, a short second segment termed the *pedicel* and a long many jointed *flagellum*. The antennae are considered to be modifications of the second head segment.

Mouth Part. These are movably-articulated appendages surrounding the mouth. The mouth parts are *chewing* type. They include a *labrum*, a pair of *mandibles*, a pair of *maxillae, a labium* and *hypopharynx*.

(i) Labrum. It hangs downwards from the membranous posterior part of clypeus and is actually an appendage of 3rd segment. It froms the anterior wall of the preoral cavity and is called the upper lip. It bears on each side of its inner anterior surface a sclerite *torma* and two rows of gustatory setae.

(ii) Mandibles. The mandibles articulate with the genae and lie on the sides of mouth just behind the labrum. These are the appendages of 4th segment and have ball and socket attachment with the head. These are heavily sclerotised and bear 3 - teeth like structure and a small serrated molar area. The upper inner margin bears a membranous lobe, the *prostheca* from which sensory hairs arise. Each mandible is worked by four muscles.

(iii) Maxillae. Two maxillae lie below the mandibles. They are regarded the appendages of 5th segment. Each maxilla is biramous appendage consists of three parts: The basal *protopodite*, inner *endopodite* and outer *exopodite*. The protopodite is made of two joints proximal *cardo* and distal *stips*. The exopodite is a long structure arising from a small projection, the *palpifer*, on the outer side of the distal end of stipes. It is formed of 5 joints and known as *maxillary palp*. The endopodite is attached to the distal end of the stipes internal to maxillary palp. It is made up of two pieces, outer *galea* and inner *lacinia*. The lacinia bears bristles along the inner margin and ends in two sharp claws. The maxillae hold food by the claws of lacinia and bring it to the mandibles for mastication. They are also used for cleaning of antennae, palps and front legs.

(iv) Labium or second maxillae. The labium lies behind the first pair of maxillae. It is formed by the fusion of second pair of maxillae along the middle line. The basal joint of the fused protopodites, is the *submentum*,

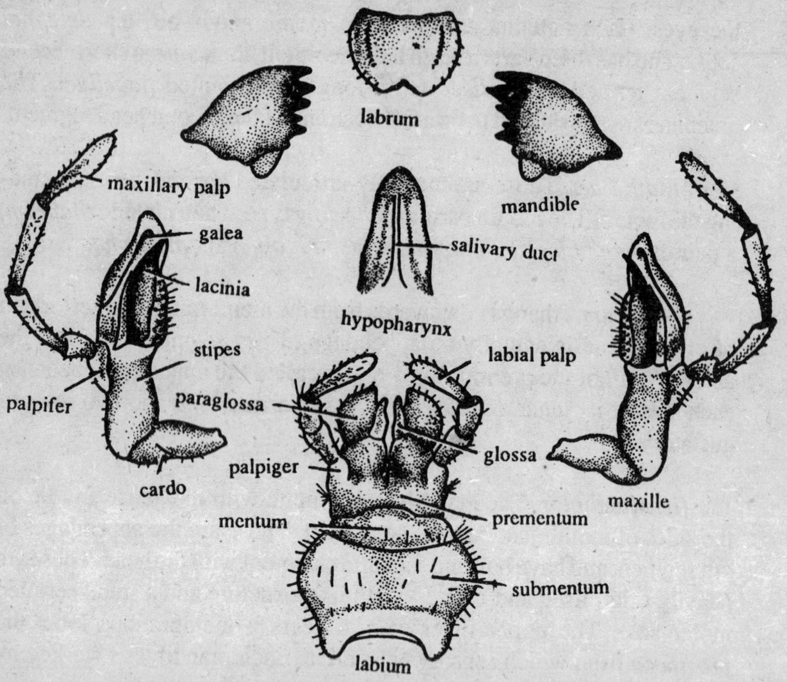

Fig 5.3 *Periplaneta* Mouth parts

the second is *mentum* and the third is *prementum.* The endopodite of labium are almost separate, and are together known as the *ligula.* Each half of ligula comprises two pieces the inner smaller *glossa* and outer *paraglossa.* The exopodites are also separate. They are called *labial palps.* Each labial palp arises from a short projection, the *palpiger*, of the premantum and consists of three joints. The labial palps bear sensory bristles. The labium takes no part in feeding, but the glossae and paraglossae together termed the *ligula* prevent the loss of food particles from the mandible thus acts as lower lip.

(v) Hypopharynx. It is the tongue or *lingua* which lies between the maxillae and labium inside the mouth cavity. The salivary duct opens at the base of the hypopharynx.

Nack. A narrow, elastic and a short neck joins the head with the thorax. Its exoskeleton is made up of chitinous plates called *cervical* plates. It is nearly an extension of an articular membrane and not a segmental region of the body.

2.*Thorax*. The thorax consists of three segments (i) anterior *prethorax,* (ii) middle *mesothorax* and (iii) posterior *metathorax*. The exoskeleton of each segment consists of a *notum* or *tergum* on the dorsal side, a *sternum* on the ventral side and a *pleuron* on either lateral side. The tergum (pronotum) of the prothorax is the largest and projects forwards to conceal the neck and a part of head also. The mesotergum and metatergum are smaller and rectangular. The sclerites of each segment and also those of the adjacent segments are joined by thin, soft, flexible *arthrodial membranes* to permit movement.

Legs. Thorax bears three pairs of legs, a pair per segment. All the legs are similar in structure. They are called *prolegs, mesolegs,* and *,metalegs* on the basis of the segment which bears them. Each leg bears five segments:

(i) *Coxa* which is attached with the thorax between pleuron and sternum, forms a morable joint.

(ii) *Trochanter,* which is smaller than coxa, can move freely on it.

(iii) *Femur*, which is long and broad.

(iv) *Tibia,* which is long and of uniform thickness.

(v) *Tarsus,* which is the last segment of the leg. It is made up of 5 movable podomeres. The last podomere ending into two small-hooked *claws* is often called *pretarsus*. Between the claws, there is a delicate, hair-covered porous pad, the *pulvillus* or *arolium.* The tarsal segments bear many fine hairs, while at their lower edges are soft adhesive pads, the *plantulae*. The pads afford grip on the smooth and slippery surface where claws fail to hold on.

Wings. Besides the legs, the thorax also bears two pairs of wings. According to their position, the wings are called the mesothoracic and metathoracic. The wings are formed as lateral extensions of the integument between the tergum and the pleuron near the anterior end of the segment. The first pair of wings are heavily sclerotised and are known as *elytra* or *tegmina*, they are protective and cover the hind wings in a folded state, the left tegmen partly overlap the right. The second pair of wings are membranous and larger, they lie folded below the tegmina

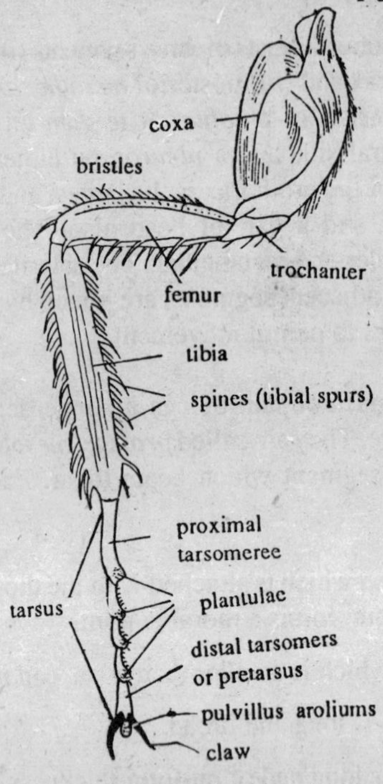

Fig 5.4 *Periplaneta* A leg

when at rest. Each wing has two thin membranous layers tubular and chitinous prolongations of the haemocoel between them knwon as *veins* or *nervures*. They are highly sclerotized and enclose small trachae and even nerve fibres.

The movements of each wing during flight are effected by two sets of muscular, the *direct* and the *indirect* muscles. The direct muscles are so called because they are inserted directly on the sclerites at the base of the wings. The indirect muscles are not inserted on the wings. One group of indirect muscle runs dorsally and longitudinally between the mesothorax and metathorax, the other group runs dorso-ventrally between the tergum and sternum.

3. *Abdomen*. The abdomen consists of 10 segments in adult but 11 segments in embryo. Typically, each segment has its own tergum and sternum. The 8th and 9th terga of female and only 9th tergum of male are largely covered by the 7th tergum. The 10th tergum projects backwards

beyond the abdomen and is cleft into two lobes whereas all abdominal segments have their own terga, there are only nine sternal plates. First sternum is small but all others are well developed. In the female the 8th and 9th sterna are invaginated and completely concealed by the enlarged boat shaped 7th sternum. This large sternum encloses a big space called *oothecal chamber* or *gynovalvular chamber,* where ootheca is formed. In males the 8th and 9th sterna are concealed by 7th sternum but there is no such chamber, however, there are a pair of small, thin and unjointed *anal styles* projecting back from the 7th sternum. Anal styles are absent in female. Other paired outgrowths are called *anal cerci* arise from the 10th tergum. These cerci are 15 jointed, longer and thicker structures. These are present in both the sexes.

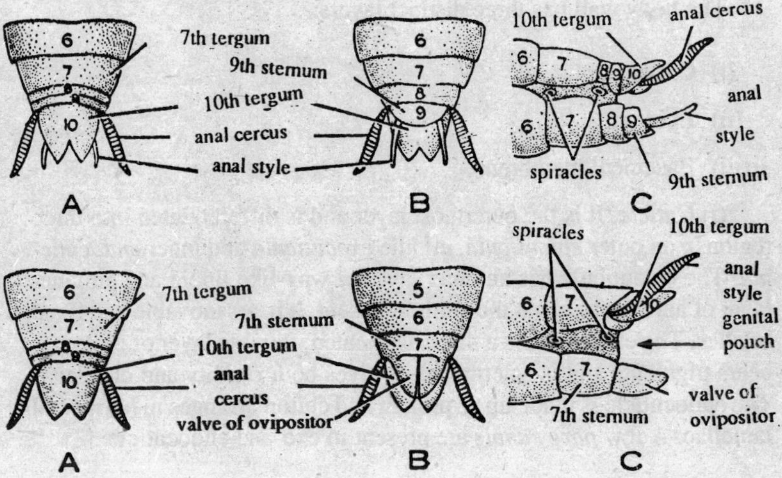

Fig 5.5 ***Periplaneta*** **Abdomen. Upper A,B,C= male, lower ABC= Female**

In made there are a group of chitinous hooks and plates near the end of the abdomen, these are the *gonapophyses* which help in copulation. Some chitinuous structures in the female have a similar position and form an *ovipositor*.

In the 10th segment just below the tergum is *anus* supported by a pair of *podical plates*. The *gonopore* is both the sexes occurs below the anus and is surrounded by gonapophyses.

Spiracles. There are 10 pairs of small slit-like apertures on the lateral sides of the body, called *spiracles.* Two pairs of these are larger and situated on the thorax, one between the pro- and mesothorax and other between meso - and metathorax. Rest of the 8 pairs are called abdominal spiracles present on the sides of abdominal segments 2-8, each being located between the two adjacent terga and two adjacent sterna.

Stink glands. The intersegmental membrane between the 5th and 6th abdominal terga is depressed on either side to form a stink gland. These glands produce a secretion with a characteristic odour.

Body wall

The body wall has three distinct layers:

(i) Cuticle

(ii) Epidermis

(iii) Basement membrane.

(i) *Cuticle.* It is the outermost layer and is differentiated into three regions : an outer *epicuticula,* middle *exocuticula* and inner *endocuticula.* The epicuticula has an outer layer of wax-like lipoid and an inner layer of hard protein, but the chitin is absent. It bears movable and fixed bristles. The exocuticle is a thick, laminated, flexible layer of chiton. It bears pigments and is sclerotized. It gives both rigidity and elasticity. The endocuticle is made up of protein and chiton arranges in horizontal lamellae. A few *pore canals* are present in exo-and endocuticle.

(ii) *Epidermis or hypodermis.* It comprises a single layer of columnar epithelial cells resting on the basement membrane. It secretes cuticle. The gland cells witch secrete movable hair-like setae that project from the surface of cuticle, are present in epidermis.

In adult cockroach large oval cells are present below the epidermal cells called *oenocytes.* The oenocytes probably secrete wax.

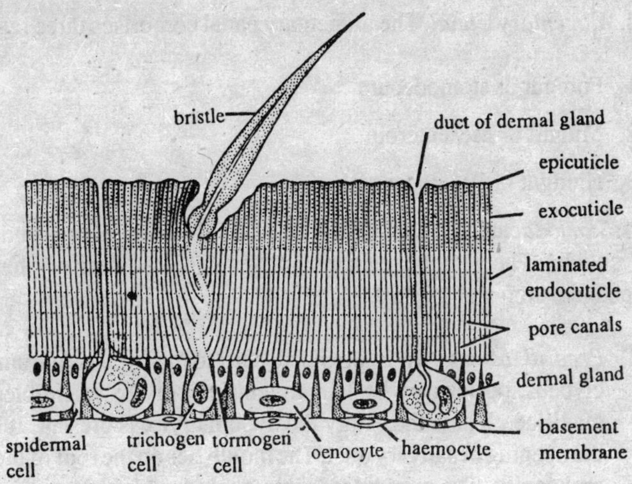

Fig 5.6 *Periplaneta* T.S. of body wall

(iii) *Basement membrane*. The basement membrane is generally 0.5 micrometer or less in thickness and in electron microscope appears as a continuous, amorphous granular layer. It is composed of mucopolysaccharides.

Endoskeleton.

The endoskeleton in cockroach is made up of chitin. The endoskeleton of head is known as *tentorium*. It lodges in the posterior part of the head. It consists of a chitinous plate with a small apertures in it. Three pairs of processes arise from this plate; *anterior arms* attached with clypeus and frons, *posterior arms* attached with the posterior part of the head and *dorsal arms* bent towards the antennae.

Body cavity

The true coelom is reduced and is represented by the cavities of gonads. The body cavity is filled with blood and is called the *haemocoel*. Loose tissues known as fat bodies or corpora adiposa are found in the body cavity.

Digestive system

The digestive system comprises the *alimentary canal* and *digestive glands* associated with it.

I. *Alimentary Canal.* The alimentary canal comprises three regions.

1. Foregut or stomodacum
2. Midgut or Mesenteron
3. Hindgut or Proctodaeum.

1. *Stomodacum.* Its inner wall is made up of epithelial cells and lines with cuticle. The stomodacum comprises: preoral cavity, pharynx, oesophagus, crop and gizzard.

(i) *Preoral cavity* . It is bounded anteriosly by the labrum and clypeus, posteriosly by labium and laterally by mandibles and máxillae. Inside the cavity the hypopharynx is present. It bears the opening of salivary duct. The mouth lies on the roof of the pre-oral cavity. The mouth leads into a tubular *pharynx.*

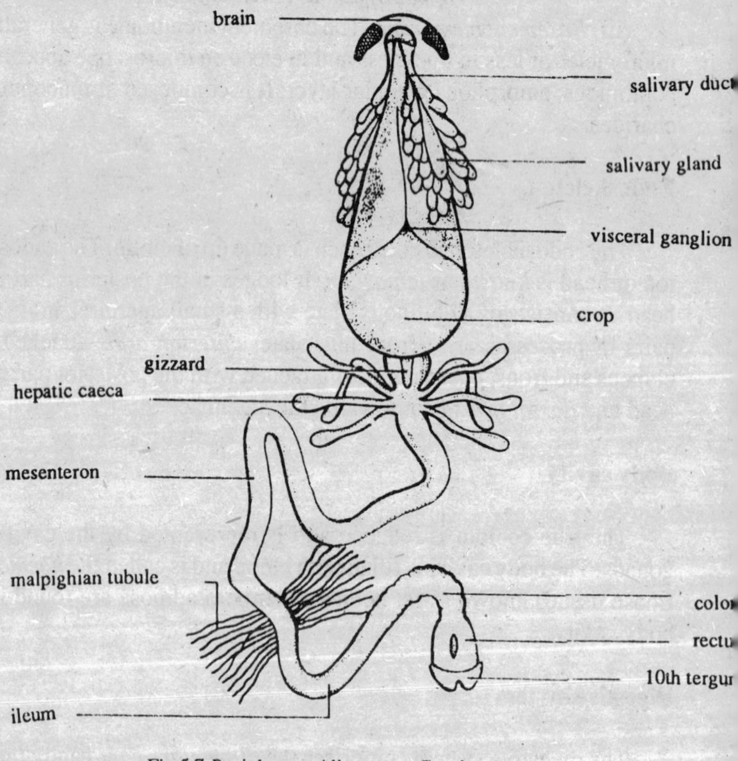

Fig 5.7 *Periplaneta* Alimentary Canal

(ii) *Pharynx.* It is a vertical canal that rises up in the head capsule and passes over the tentorium, and then bends to open into the oesophagus near the occipital foramen.

(iii) *Oesophagus.* It is a narrow tube that is present in the neck and extends into thorax to open in the crop.

(iv) *Crop.* The crop is a large thin walled, pear-shaped sac extending well into the abdomen. It serves as a reservior for storing food. Its outer surface is covered by a network of tracheae. Its internal epithelial and cuticular lining is very much folded.

(v) *Gizzard or proventriculus.* The crop leads behind into a small, thick walked, muscular and cone shaped proventriculus. Anteriorly it bears an *armarium* and posteriosly a *stomodaeal valve.*

The armarium possesses internally six longitudinal folds that greatly reduce its lumen. These longitudinal folds alternate with six longitudinal grooves. The cuticular lining of each longitudinal fold forms (a) in the anterior part, a *thick plate* produced centrally into strong, sharp teeth and (b) in the posterior part, a *thin plate.* Behind each longitudinal fold the cuticular lining of the gizzard forms a soft cushion-like lobe, the *pad* or *pulvillus,*

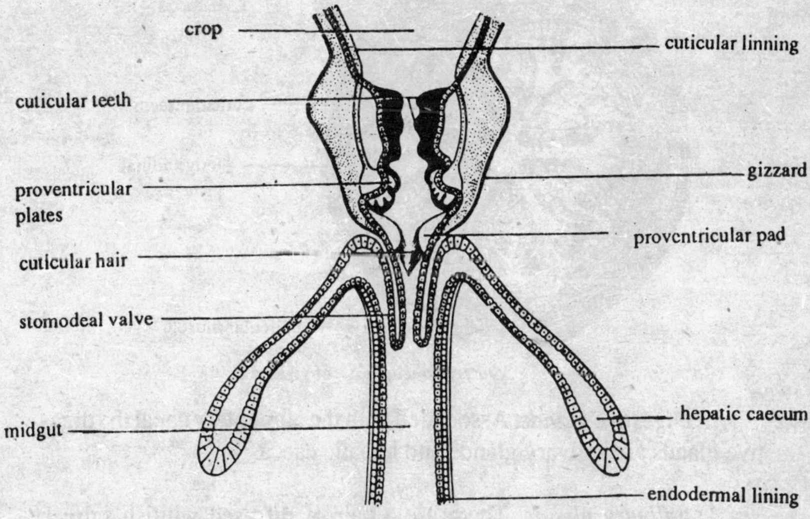

Fig 5.8 *Periplaneta.* Sectionn through gizzard and mesenteron

with long, backwardly directed hair. Behind the region of the pads the proventriculus extends into the lumen of the midgut as the stomodaeal valve. The latter folds back on itself, and is thus double-walled.

2. *Mesenteron*. Behind the gizzard is a tubular mesenteron or midgut lined with endodermal cells, it performs the tasks of completing the digestive process and absorption of food. The endodermal cells of the midgut break down while secretory and their contents are discharged into the lumen and fresh cells are formed. From the anterior end of the midgut arise eight tubular *hepatic caeca*. From its posterior end 70-80 very fine, thread-like yellow *Malpighian tubules* are given out.

3 *Proctodaeum.* Its wall is also made up of ectodermal epithelium and lined internally with cuticle. It consists of three parts ilium, colon and rectum. At the junction of mesenteron and ilium are present Malpighian tubules arranged in six groups. The inner surface of rectum is raised into six longitudinal ridges, the *rectal papillae* which help in the absorption of water. The rectum opens out through the anus.

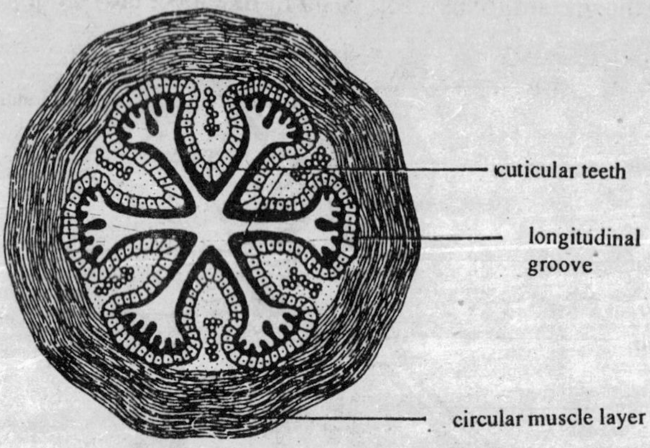

Fig 5.9 *Periplaneta*. T.S. of gizzard

II. *Digestive glands*. Associated with the alimentary canal the digestive glands are salivary glands and hepatic caeca.

1. *Salivary glands.* There are a pair of diffused whitish salivary glands, one on either side of crop in the thorax. Each gland has two

branching glandular parts whose ducts form a common glandular duct. Between two branches of each gland lies a thin walled sac-like reservoir. The ducts of both the reservoirs unite to form common reservoir duct which opens into the hypophanynx. The common glandular duct opens into the common reservoir duct.

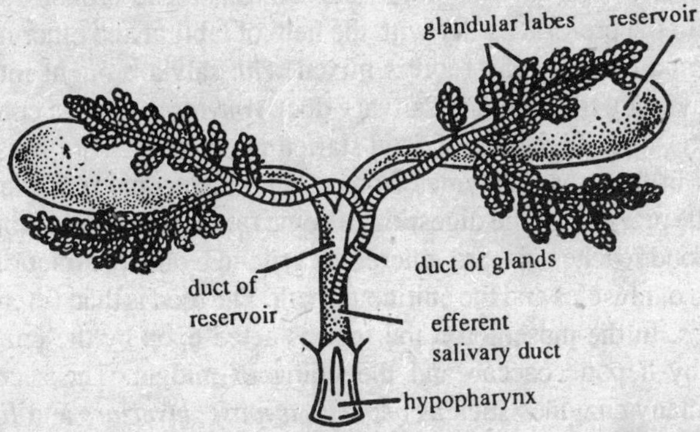

Fig 5.10 *Periplaneta.* Salivary glands

Each gland consists of several secreting *acini* or *lobules* in grape-like clusters and connected together by fine tubules. Two types of cells are found in acini, *zymogenic cells* and *ductule containing cells.* Both types of cells secrete saliva which contain enzyme zymase and mucoid sustance. *Intercalary ducts* from the various acini join each other and are connected to the reservoir where saliva is stored.

(2) *Hepatic caecae.* These are 7-8 finger-like, hollow tubes, ending blindly and open into the midgut. They secrete digestive enzymes.

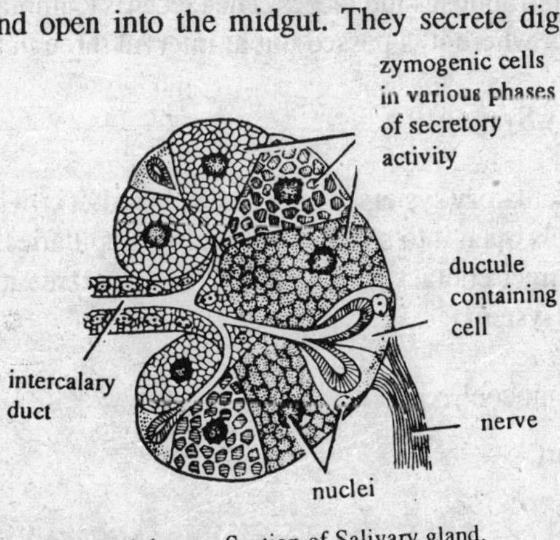

Fig 5.11 *Periplaneta.* Section of Salivary gland.

Food and feeding

The cockroach is an omnivorous animal and feeds on all sorts of food except metals, bricks, stones etc. The food is located by the sense of small, whose receptors lie on the antennae and palps. The food is caught and masticated with its mandibles and maxillae. The broken food is pushed into the pre-oral cavity with the help of labrum and other mouth parts. During chewing the food is mixed with salvia brought into the preoral cavity by the common salivary duct. *Amylase* of saliva converts starch into sugar. Digestion of food started in pre-oral cavity by saliva continues in the crop. Enzymes from midgut also come into the crop through the gizzard for the digestion of some fats and proteins. From the crop the food reaches gizzard where it is ground by the contraction and expension of muscles and the chitinous teeth. The food is than filtered by the bristles. In the mesenteron the food is acted upon by the enzymes secreted by hepatic caecae and the lining of midgut. The secretion contains many enzymes such as *peptase, trypase, invertase* and *lipase*. The digested food is absorbed by the lining of midgut and hepatic caecae.

In the midgut, *peritrophic membrance* is formed round the food for the protection of the lining of it from the abrassion by hard indigestible particles in the food. The membrance is a thin chitinous tube formed of the secretion from the funnel-like extension of the gizzard round the food. It is permeable to both, the enzyme and digested food.

The undigested food reaches the rectum through hindgut. The absorption of water takes place in the rectum and the undigested food is converted into almost solid faeces. The faeces is temporarily stored in the rectum from where it is passed out at interval through the anus.

Circulatory System

The circulatory system is poorly developed. It is of *open type* i.e. the blood vessels open into spaces and not into capillaries. Thus the blood comes in direct contact with tissue. There are three main parts in the circulatory system.

1. Haemocoel.
2. Heart.

3. Blood.

1. *Haemocoel.* The haemocoel devoid of true epithelial lining. It is filled with haemolymph. It is partitioned by a dorsal and a ventral diaphragm into three sinuses - (i) the dorsal *pericardial sinus,* containing the *drosal blood vessel* or *heart,(ii)* the middle *preivisceral sinus,* containing the gut, and (iii) the ventral *perineural* or *sternal sinus,* containing the nerve cord. The diaphragms or septa are provided with pores or *fenestrae.* The ventral diaphragm also extends as septa into the legs. In each leg the septum divides its cavity into two sinues, one for the outward and the other for inward flow of blood.

The pericardial septum is slightly bulged upwards. There are a series of paired triangular muscles, the *alary muscles* attached to the pericardial septum by their broad bases and to the terga by their pointed ends. There are 12 pairs of alary muscles in *Periplaneta.;*

2.*Heart.* The heart is a long, muscular contractile tube extending along the middorsal line just beneath the terga of thorax and abdomen. It is enclosed in the pericardial sinus. It is composed of 13 chambers, each

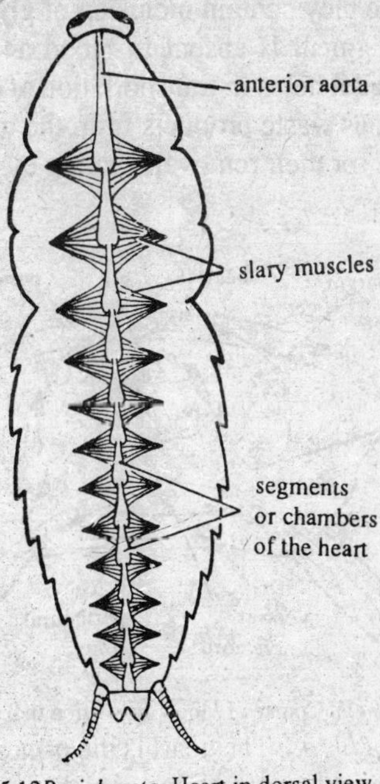

Fig 5.12*Periplaneta.* Heart in dorsal view

chamber communicating by a valvular opening (*auricular valve)* with the one in front of it. The chambers communicate to the pericardial sinus by paired lateral openings called *incurrent ostia.* Ostia are also valvular allowing the entry of blood from the pericardial sinus into the heart. The heart is closed behind and is continued infront as a short tube, the *aorta* which opens into the haemocoel of the head. It is provided with *aortic valves.*

Nutting (1951) has described that in roach' *Blaberus,* there are 6 paired lateral segmental blood vessels of which 2 thoracic and 4 abdominal pairs. Through these blood vessels the blood leave the heart.

3. *Blood*. The blood or haemolymph is contained in the haem ocoel. It is only the extracellular fluid and makes 15-75% volume of the insect.

It consists of liquid plasma and numerous colourless blood cells or *haemocytes*. There arc about seven types of haemocytes (*Wigglesworth,* 1965). The principal function of the haemocytes is *phagocytosis* i.e. the ingestion of small solid particles. Some haemocytes fulfil the function of food storage since they contain inclusion of glycogen and fat. In insect the respiratory pigment is absent in blood and thus plays no role in respiration. Instead it helps in transportation of digested food materials, collects nitrogenous waste products from the tissues and brings to the extractory organs for their removal, healing of wound transportation of the hormones etc.

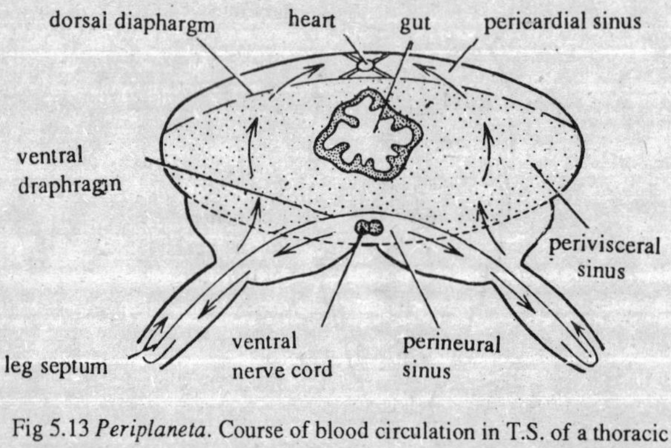

Fig 5.13 *Periplaneta.* Course of blood circulation in T.S. of a thoracic segment

Circulation of blood. The heart is the principal pulsatory organ and undergoes rhythmical contractions which are brought about by the muscle fibres of its walls. The rhythum of the heart appears to be purely myogenic (*Miller*, 1974) though it is influenced by nervous system. Contraction of the heart takes the form of a wave of peristalsis which runs forward from the posterior end. The blood is poured into the head sinus through the dorsal aorta. These is an *accessory heart* at the base of each antenna which send haemolymph into the antenna. From the head sinus, the haemolymph flows into the periviseral and sternal sinuses. Now the alary muscles contract, so that the pericardial septum becomes Flat. This increases the capacity of pericardial sinus. The haemohymph, therefore, flows from perivisercal sinus to the pericardial sinus through the perforations of diaphragm. On relaxation of the alary muscles, the pericardial diaphragm bulges upwards, this presses blood which enters the heart through ostia. The rate of heart beat in *Periplaneta* in 49 per minutes.

Respiratory System

In the cockroach respiration takes place by means of internal air-tubes known as *tracheae*. These ramify through the organ of the body and its appendages, the finest branches being called *trachioles*. The air generally enters the tracheae through *spiracles*, which are lateral in position in the thoracic and abdominal region.

1.*Spiracles*. The spiracles are derives from the mouths of the ectodermal invagination which give rise to the *tracheal system*. Ten pairs of spFig 5.10 *Periplaneta*. Salivary glandsiracles are found in *periplaneta* 2- pairs in thorax and 8 pairs in abdomen. The first pair lodge between prothorax and mesothorax, second pair between mesothorax and metathorax and

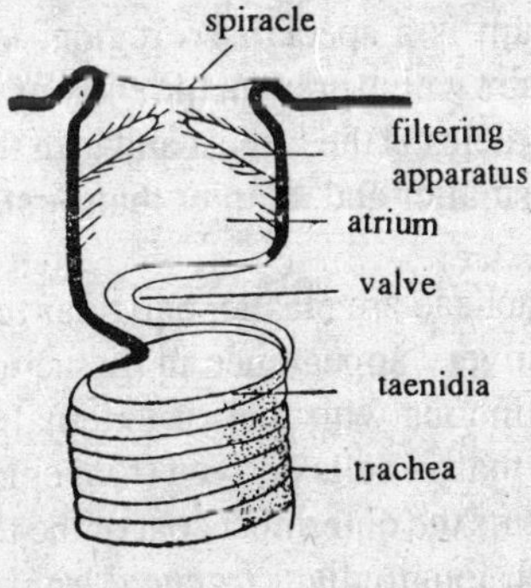

Fig 5.14 *Periplaneta*. A spiracle with filtering apparatus, valve and atrium

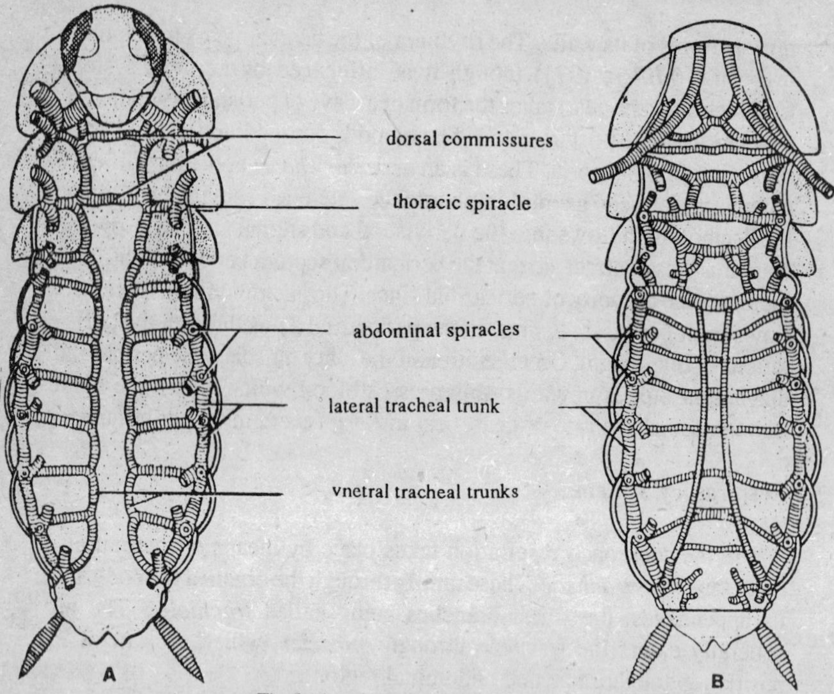

Fig 5.15 *Periplaneta.* Salivary glands

remaining 8 pairs are present between the sternum and pleuron of first to eighth abdominal segments. Each spiracle is surrounded by an annular sclerite called *peritreme* and leads into a chamber called *atrium* from which the trachea arises. The spiracle is providided the *closing apparatus.* The apparatus consists of one or more muscles with associated cuticular parts and by closing the spiracular aperture, prevents excessive loss of water vapour. The atrium is a specialised region, lacking taenidia or intima, provided with hairs which prevent the entry of sand particles with air. The first pair of spiracles is the largest and is in the form of oblique slit. The second pair is smaller and simpler than first.

2.*Tracheae*. The tracheae are elastic, cuticular tubes, which, when filled with air, have a silvery appearence in dissections. The cuticular lining or *intima* is continuous with that of external body surface and secreted by the tracheal matrix cells derived from epidermis. A delicate basement membrane forms the outermost coat of the tracheae. There are 3 pairs of large, parallel, longitudianl *tracheal trunks*, one dorsal, one ventral and one lateral in position. These are connected together by *transverse commessures.*

The tracheae branch and rebranch to from ultimate finer branches called *tracheoles.* They have a diameter of only i.e. Their cavities are intracellular, i.e. each tracheole is made of a single cell. Their walls are very thin and devoid of intima, instead they are lined by a protein, the *trachein*. They are permedable to water and are usually filled with a fluid in which oxygen dissolves and diffuse to the tissue.

Mechanism of respiration

By the contraction and expension of *tergosternal muscles* the terga and sterna contract and expand as a result of which inspiration and expiration takes place. Expiration is an active process where as inspirations is passive. The first pairs of thoracic and abdominal spiracles, always remain open but the remaining open during inspiration but close during expiration. Air enters the spiracles during inspiration and comes

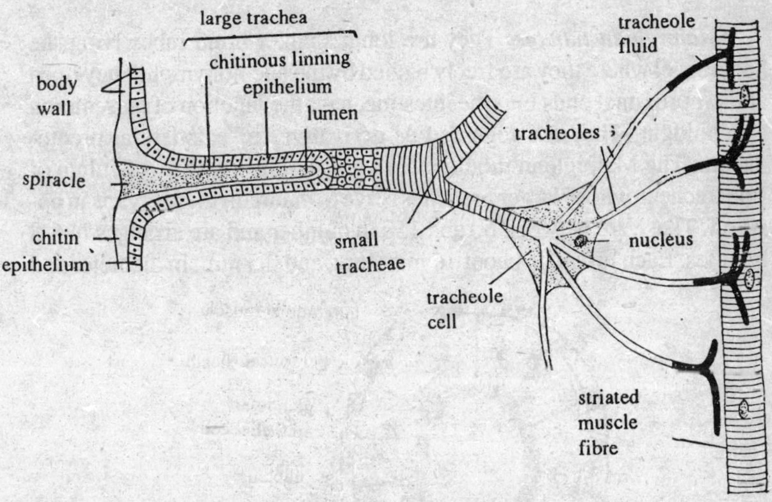

Fig 5.16 *Periplaneta.* A Trachea

to the tracheae, then it comes to the tracheoles which contain fluids, the oxygen dissolves in these fluids and reaches the cells of the tissues. Opening of the spiracles and subsequent diffusion of air arc due to the stimulation of spiracles by carbon dioxide. In expiration some carbon dioxide may pass out through the spiracles but the major part of it diffuses out through the cuticular covering of the body. Carbon dioxide also dissolves in the plasma and reaches the body surface which is permeable to gases and allows Carbon dioxide to pass out.

When the cockroach is running, the oxygen requirement increases and speical respiratory movements of the abdomen start. The fluid present in the tracheoles is withdrawn into the body cells and the air rushes forward to take its place, the oxygen reaches itself directly to the cells of the body tissues. It is not supplied through the medium of blood or haemolymph.

Excretory System

The excretory system is simple in cockroach. The principal excretory organs are the *Malpighian tubules,* working in conjunctions with the hindgut. Some haemocoelomic structures, such as the nephrocytes, fat bodies etc. have also been regarded in the past as excretory organs.

Malpighian tubules. They are long, slender blind tubes lying the haemocoel where they are freely bathed by the haemohymph. They open at their proximal ends into the intestine, near the junction of mesenteron and hindgut. They are regarded as derivatives of ectodermal proctodaeum. The Malpighian tubules are richly supplied with a reticulum of fine tracheae whose larger branches serve to maintain these organs in position. These are between 60 and 150 in number and are arranges in 6-8 bundles. Each tubule is baout 16 mm. long and 0.5 mm. in diameter and

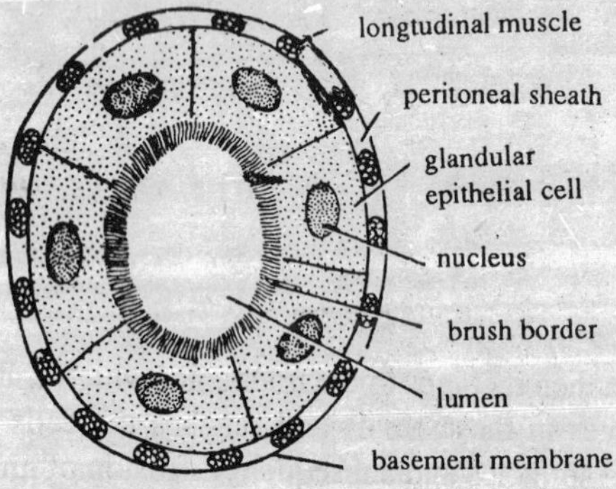

Fig 5.17 *Periplaneta.* T.S. of Malpighian Tubule

is lines by glandular epithelium with a characteristic *brush border*. The glandular cells extract the uric acid, water and inorganic salts from the haemolymph and secrete them into the lumen from where they flow into the hind gut to be eleminated with the faeces. The reabsorption of water takes place in the rectum. The Malpighian tubules have been observed to show peristaltic movements that bring more and more haemolymph in their contact.

The role of Malpighian tubules in regulating water balance and ionic concentrations presents complex biophysical problems still not fully elucidated.

Fat bodies. The fat-body is composed of irregular masses or lobes of rounded or polyhedral cells. There are usually vacuolated and contain inclusions of various kinds. The fat bodes are derived from mesoderm by differentiation. The nitrogenous waste products (uric acid and urates) are present in the form of excretory granules in specialized *urate cells* of the fat body. Throughout the life these products are deposited in these cells.

Besides urate cells the other types of cells present in the fat body are : (i) *Trophocytes* which store fat, glycogen and proteins as reserve food materials.(ii) *Oenocytes* which are believed to secrete cuticular lipids and it has been suggested that they may be the site of ecdysone production and (iii) *Mycetocytes* which are filled with symboint micro-organisms that help in amino acid synthesis.

Uricose glands. The mushroom glands of male cockroach possesses long, blind tubules at its periphery called *uricose glands*. These gland store uric acid and discharge it over the spermatophore during copulation.

Cuticle. some excretory substances probably accumulate in cuticle of the body wall shed along it during *ecdysis* or *moulting*.

Nervous System

A well developed nervous system is present in cockroach. It comprises three main parts:

I. Central nervous system

II. Peripheral nervous system

III. Autonomic nervous system

1. *Central nervous system*: It comprises *brain* or *cerebral ganglian, suboesophageal ganglion, circum oesophageal connectives* and *nerve cord.*

The *brain* or *supra-oesophageal ganglion* is a large and bilobed mass located in the head above the oesophagus almost between the bases

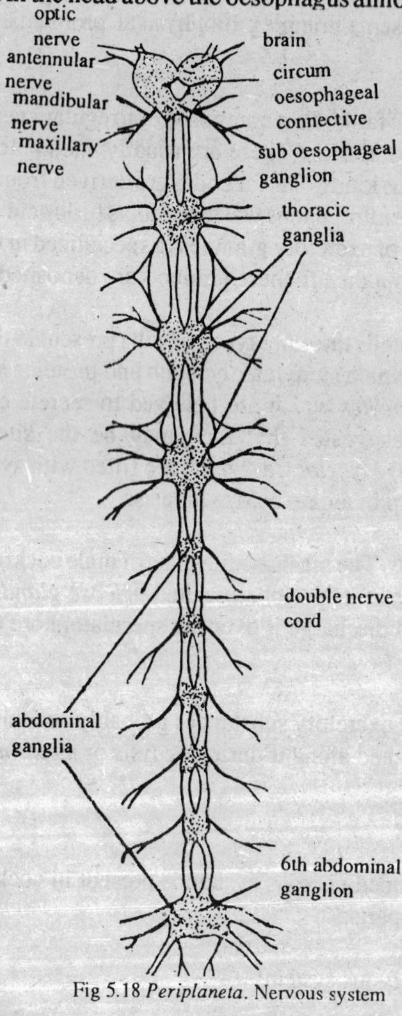

Fig 5.18 *Periplaneta*. Nervous system

of the anatennae. It represents three fused ganglia of the head region. It is divided into three parts - the *protocerebrum, deuterocerebrum* and *tritocerebrum.* The remaining three ganglia of the head fuse to form the *sub-oesophageal ganglion* which lies below the oesophagus, posterior to the brain. The brian and the sub-oesophageal ganglion are connected together, on either side, by a *circum-oesophageal* connectives

From the sub-oesophageal ganglion, a double *ventral nerve cord* runs posteriorly along the mid-ventral line of the thorax and abdomen. The nerve cord bears nine ganglia, three in the thorax and six in the

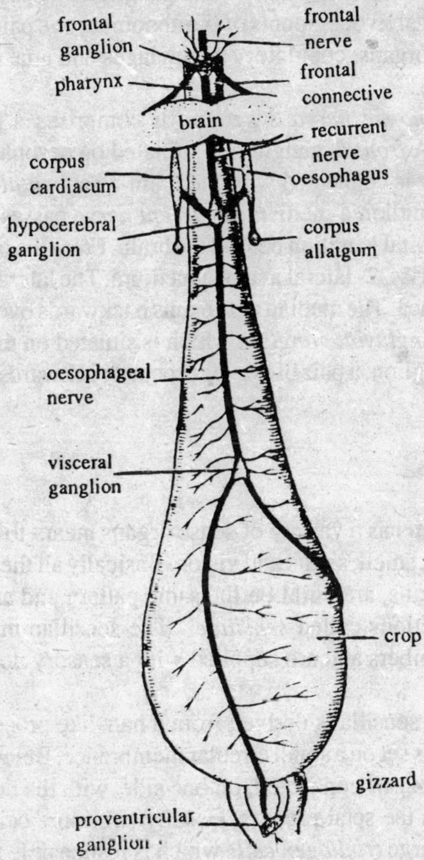

Fig 5.19 *Periplaneta.* Sympathetic Nervous System

abdomen. The last abdominal ganglion is the largest, representing many fused ganglia of the posterior abdominal segments.

II. *Peripheral nervous system.* From the central nervous system arise nerves which go to different parts and constitute the peripheral nervous system. From the brain three pairs of nerves arise *optic, antennary and labro frontal* nerves and go to the eyes, antennae and labrum respectively. From the sub-oesophageal ganglion also three pairs of nerves arise and go to the mandibles, maxillae, and labium as *mandibular, maxillary* and *labial* nerves. Several pairs of nerves arise from each ganglion of the nerve cord and they innervate the various parts of their segment; but from the last abdominal ganglion five pairs of nerves arise and go to the last five segments of the abdomen, one pair to each segment reproductive organs, copulatory appendages and anal circi.

III. *Antonomic nervous system.* It comprises a few ganglia and connectives. A *frontal* ganglion is situtated on oesophages just in front of the brain. It is connected with the brain by a *frontal connective* from the frontal ganglion a mediam *recurrent nerve* passes beackwards and bears the occipital ganglion behind the brain. From the occipital ganglion three nerves arise, 2- lateral and one medium. The lateral nerves meet the corpora cardiaca. The median nerve runs backwards over the oesophagus and joins the *ingluvial ganglion* which is situated on the crop. From the ingluivial ganglion, a pair of nerves proceed backwards over the alimentary canal.

Sense Organs

Cockroach has a variety of sense organs meant to catch the stimuli of tocuh, taste, smell, sound and vision. Basically all these organs, except the visual organs, are build on the same pattern and are in the form of minute projections called *sensillae.* The sensillae may be present in enormous numbers and are supplied with a sensory nerve fibre.

A typical sensilla is derived from a hair-like process of the cuticle or seta which is set on a small circular membrance. Below this membrane is a large *sense cell* connected, on one side, with the nerve fibre and on the other with the setal process. Usually two more cells are associated with seta. A large *trichogen cell,* which is responsible for secreting seta and a *tormogen cell* which gives rise to the membrane. The typical sensilla is modifies into two types - the *tactile sensillae* in which there is a single sense cell and the *chemical sensillae* in which there is a group of such cells. In the latter, the seta itself is short and permeable to the fluids or vapours. The *acoustic sensillae* are so modified as to be present in groups and associated with certain membranes which vibrate by sound waves.

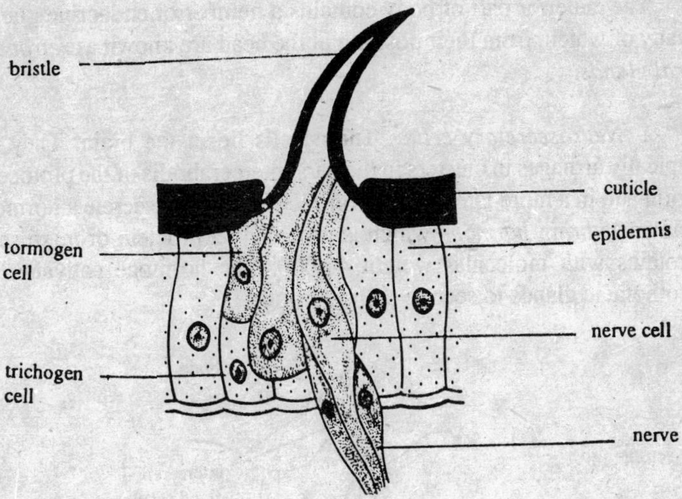

Fig 5.20 *Periplaneta.* A Sensilla

In *Periplaneta,* tectile sensillae are distributed over the general body surface but specially on antennae, palpi, anal cerci and distal segments of the legs; olfactory sensillae are present on the antennae; gustatory sensillae on the palpi and other mouth parts and the legs; and acoustic sensillae on the base of the abdomen.

*Compound eyes.*Of all the organs of special sense, the most important is a pair of large, sessile, *compound eyes* in the form of black, kidney-shaped patches on the sides of the head capsule. Each consists of about 2,000 visual elements or units, called *ommatidia,* similar in structure to those already described for the prawn in Chapter 2

The pigment cells are, however, not retractile in *Periplaneta* so that it can see only by apposition images.

Ocelli. At the base of each antenna there is a fenestra which represents an ocellus or simple eye. Each comprises a single lens and is capable of preceiving only the light and not the image formation.

Endocrine Glands

The anterior part of body contains a number of endocrine glands, many of which, from their position in the head are known as *retrocerebral glands.*

1. *Neurosecretory cells* . These cells lie in the brain. They are typically arrnages in clusters in the pars intercerabralis of the protocerebrum and in a more lateral group on each side. They secrete a hormone known as *brain hormone* which is probabaly a proetein or mixtuse of proteins with molecular weight 40,000. The hormone activates the prothoracic glands to secrete their hormone.

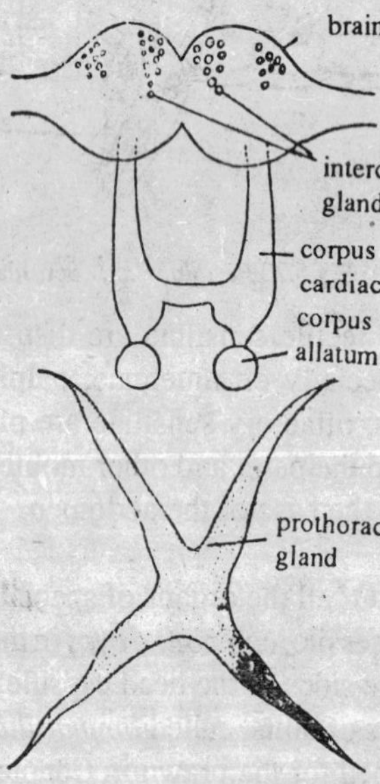

Fig 5.21 *Periplaneta*. Endocrine glands

2. *Corpora cardiaca.* Situated in close association with the aorta behind the brian is a pair of small bodies, the *corpor cardiaca* . Each is

connected to the protocerebrum by a pair of nerves and to the hypocerebral ganglion by a single nerve. The hormome of corpora cardica are those produces by its intrinsic cells and by the neurosecretory cells of the brian. The intrinsic cells are known to secrete at least two metabolic hormones, the *hyperglycemic* and *adipokinetic*.

3. *Corpora allata*. These endocrine glands are situated laterally behind the brain, usually near the corpora cardiaca. They are paired, spherical or oval bodies. The hormone secreted by these glands is known as *juvenile hormone*or or *neotenin* whose most important function is to inhibit the realization of imaginal characters during postembryonic development. During last larval or nymphal instar the corpora allata become inactive and the resulting change in hormonal balance leads to metamorphosis. In adults it exerts a gonadotróphic influence in the adult females. There is no gonadotropic effect in males but their secretion may be needed for full development of the accessory reproductive glands and for normal sexual behaviour *(Barth* and *Lester*, 1973).

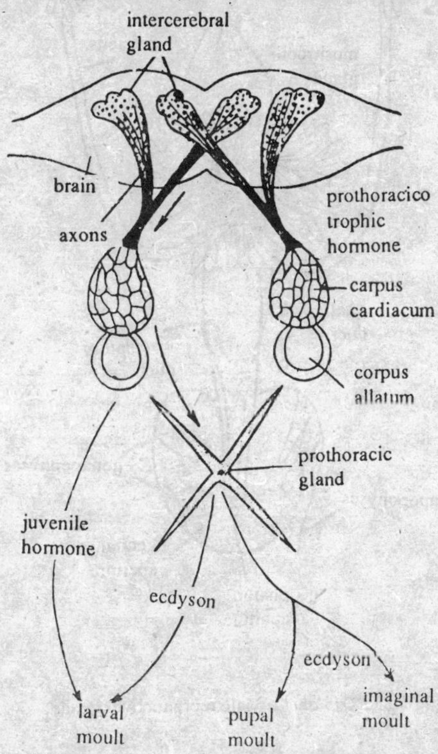

Fig 5.22 *Periplaneta*. Interaction of harmones

4. *Prothoracic glands.* These are paired structure, also known as *thoracic glands, pericardial glands, ecdysial glands* or *ventral glands.* They are fairly large and irregular glands situated in prothorax and are well supplied with tracheae and in most cases they are provided with nerves. They secrete a hormone known as *ecdyson* which controls moulting of the nymph. These glands atrophy completely in adult insect.

Reproductive System

The cockraoch is unisexual with a distinct sexual dimorphism.

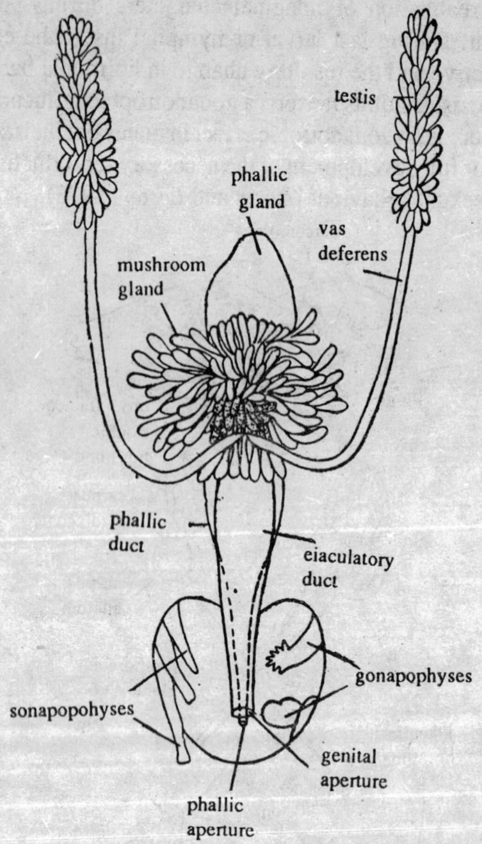

Fig 5.23 *Periplaneta.* Male reproductive organs

Male reproductive organs. The male reproductive system comprises a pair of testes, vasa deferentia, ejaculatory duct, seminal vesicles, mushromm gland and conglobate gland.

*Testes.*A pair of testes, lying embedded in the fat body, on either dorso-lateral side of the 4th to 6th abdominal segments. Each testis is a trilobed structure and consists of numerous small whitish *follicles* arranged in longitudinal series. In the adult they become nonfunctional and reduced.

Vasa deferentia. From the hind end of each testis, there starts a fine duct, the vas deferens. The two vasa deferentia run backwards and inwards to open into a wide median duct, the *ejaculatory duct.*

Ejaculatory duct. The ejaculatory duct arise in 7th segment runs backwards and opens out at the male genital aperture.

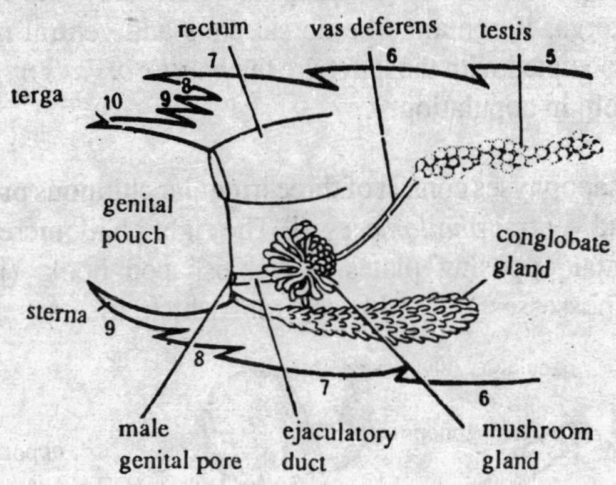

Fig 5.24 *Periplaneta.* Male reproductive organs in lateral view

Mushroom gland. It is present at the junction of vasa deferentia and ejaculatory duct. The mushroom gland or *utricular gland* consists of a number of blind finger - like diverticula opening into the anterior end of the enjaculatory duct. The diverticula are of the three types: (i) long slender tubules known as *utiriculi majores* or *peripheral tubules* arising from the anterior end (ii) short tubules, the *uticuli breviores* making up the major portion of gland and (ii) postero-median short but relatively bulbous tubules known as *seminal vesicles.*

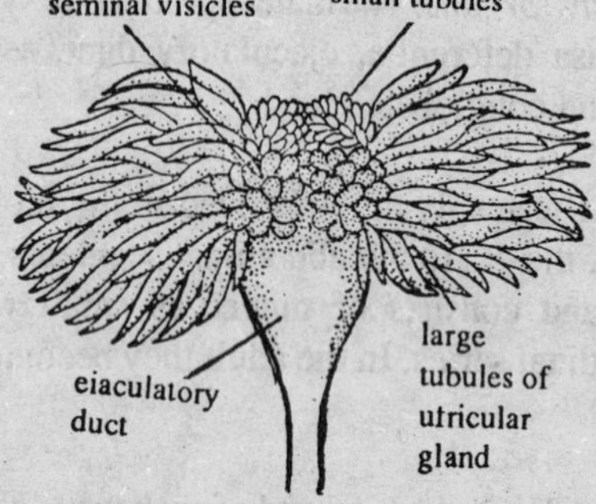

Fig 5.25 *Periplaneta.* Uricose gland

Phallic or conglobate gland. It is a large, white elongated, sac-live structure lies beneath the ejaculatory duct that opens out near the male genital aperture. Its function is unknown.

External genitalia. The genital pouch lies below the 9th and 10th abdominal terga. It contains the dorsal anus and ventral male genital opening. Associated with the latter are the *phallic organs* or *gonopophyses* which help in copulation.

The gonapophyses consist of three irregular chitinous processes, the left, right and ventral *phallomeres.* (i) The right phallomere consists of two horizontal opposing plates, a sickleshaped hook. (ii) The left phallomere possesses a broad base, from which arise several lobes or

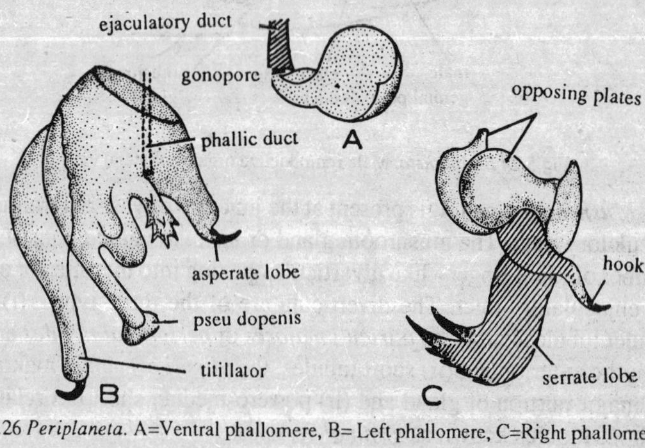

Fig 5.26 *Periplaneta.* A=Ventral phallomere, B= Left phallomere, C=Right phallomese.

arms. A long slender lobe, the *titillator*, with a curved hook, lies on its extreme left. On its right is a short lobe, the *pseudopenis*, with a broad hammer-like head. Inner to it are three short, soft lobes, one of which, called the *aspirate lobe*, bears a curves hook. (iii) The ventral phallomere is a brown broad plate without lobes. It lies close to the gonopore.

The sperms pass into the seminal vesicles and are glued together to form a *spermetophore* from the secretion of seminal vesicles. A spermatophore is pear-shaped body, about 1.3 mm in diameter and its wall composed of three non- cellular coats. The first coat is secreted by the secretion of long tubules of utricular gland and sperms flow into it. The second wall is secreted by the secretion of ejaculatory duct and the third wall is secreted by the secretion of phallic gland

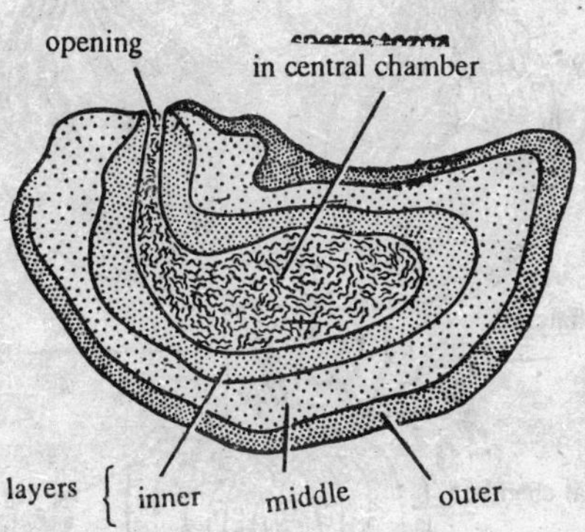

Fig 5.27 *Periplaneta.* A spermatophare

Female reproductive organs. The female reproductive organs comprises, ovaries, oviduct, vagina, gynatrium, collaterial glands and spermathecae.

Ovaries. There are a pair of ovaries lying between 2nd to 6th abdominal segments. Each ovary is made up of 8 ovarioles. Which are yellowish in colour. Each ovariole is made up of 5 parts: The anterior most thread-like structure called *terminal susponsary ligament* which is syncytical. It is followed by *germarioum* containing cells at an early stage of egg formation. It is followed by *vitellarium* containing mature ova and gives beaded appearence. It is followed by *egg chamber* which contains only one mature ovum. It is followed by *perdical oviducts*. Posteriosly all the eight ovarioles of each ovary join to form a wide oviduct.

Vagina . It is a formed by the union of both the oviducts. The vagina opens into the gynatrium by a slit like female genital pore situated in the 8th segment.

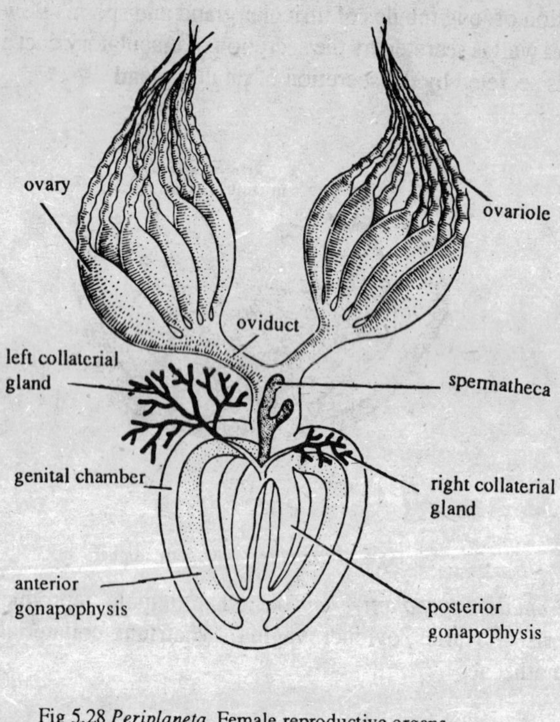

Fig 5.28 *Periplaneta*. Female reproductive organs

Spermathecae. There is a pair of spermethecae of unequal size, one spermethecae being larger than the other. The two unite to form a short common duct which opens into the genital chamber on a small spermathecal papillae. It is claimed by some workers that there is a single spermatheca and it has a lateral coiled caecum.

Collaterial glands. A pair of highly branched collaterial glands lie behind and above the ovaries. The left gland is more opaque and well developed. The right gland is transparent and less developed. The two open into *gynatrium* or *genital pouch* through their openings. The secretion of these glands provides material for the formation of *ootheca* or egg case round a group of eggs.

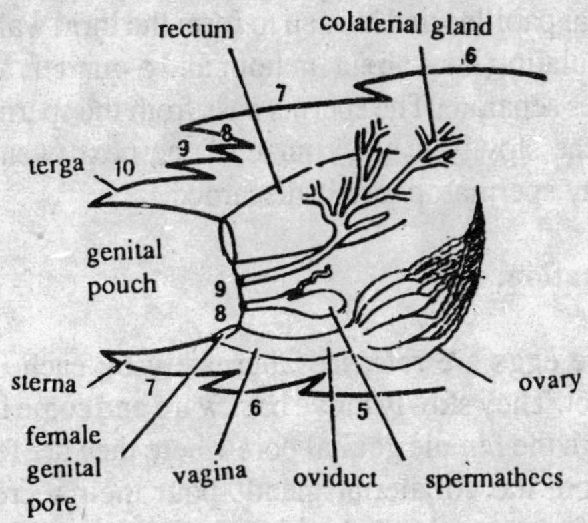

Fig. 5.29 *Periplaneta* Female reproductive organs in lateral view

External genitalia. The gynatrium or genital pouch is formed by the 7th, 8th and 9th sterna. The sternum of 7th segment is boat-shaped and splits posteriosly into two parts, the *apical lobes* or *gynovalvular plates*. This forms the floor and side walls of the brood pouch. The 8th and 9th sterna form the anterior wall and roof of brood pouch respectively. The pouch is divided into two chambers - a large posterior *oothecal chamber* and an anterior *genital chamber* into which open the vagina, spermatheca and collaterial glands.

The gonopore is surrounded by three pairs of symmetrically ar-ranges, plate-like, chitinous processes, the *gonapophyses*. Of these, one pair arises from the 8th abdominal segment, while the other two pairs belong to the 9th abdominal segment. The gonepophyses serve to hold the ova as they are laid and also aid in the formation of ootheca round them. They, thus, act as *ovipositor*.

Copulation. The breeding season of cockroach is from March to September. Copulation occurs in night. The male and female cock-roaches come opposite to each other in such a way that the posterior ends of their abdomen touch each other. The male opens the genital chamber of the female with the help of its titillator and pushes its phallomeres into the genital chamber of the female. At the same time the dorsal phaellomere opnes the male genital aperture through which the spermatophore is deposited on the spermathecal papilla. The phallic secretion is now poured over the spermatophore which fixes the spermatophore firmly to the spermathecal papillae and harden to form the thrid wall of spermatophore. The copulation lasts about an hour and a quarter, after which the two cockroaches separate. The sperms pass from the spermatophore into the spermathecae slowly in the course of the next twenty hours after which the empty spermatophore is discarded.

Ootheca Formation.

The mature eggs are released alternately by each ovary into the common oviduct. They slowly move backward and come into the genital chamber through the female genital pore where they are fertilized. Soon after fertilization, the collaterial glands pour their secretion over the fertilized eggs forming a hard covering over them called ootheca. The

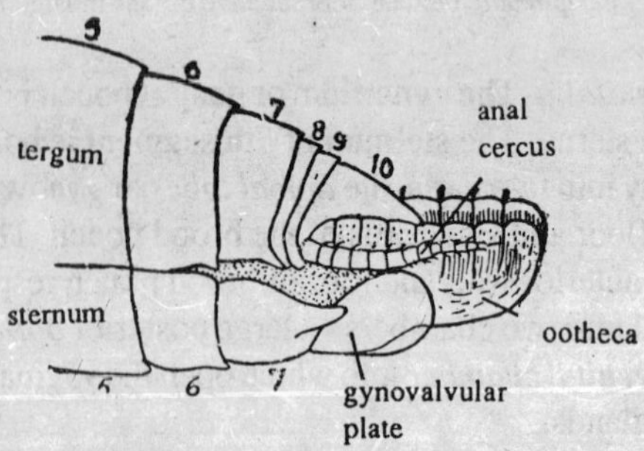

ootheca is shaped and moulded by the ovipositor and the wall of oothecal

Fig. 5.30 *Periplaneta* Ootheca protruding from abdomen of female.

chamber. The ootheca is 12 mm in length. On one side it has a straight crest with a serrated margin. It contains 16 fertilized eggs standing vertically in two rows. The female cockroach moves about with ootheca for several days and then keeps it in a dry, dark and safe place.

Development. The eggs are enclosed in ootheca. Each egg is laterally compressed being concave on one side (future ventral side) and convex on the other. Cleavage is superficial. The nucleus divides and redivied and surrounded by a stellate mass of protoplasm forming the cleavge cells. When cleavage- cells in large number are produced, they migrate to the periphery of the egg forming blastoderm, surrounding the yolk. The egg becomes a blastula, gastrulation takes place shortly after, A differentiation in the gastrula results in an early embryo which separates from the blastoderm by an amniotic cavity. Its three germinal layers gradually from the various organs of nymph.

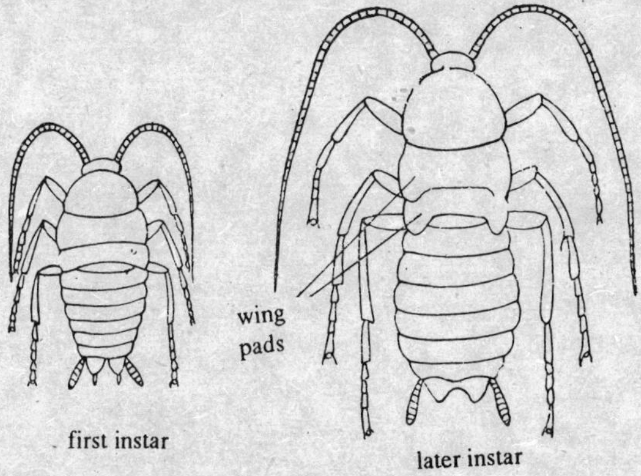

Fig. 5.31 Nymphs pf *Periplaneta* A=First instar B=Second instar

Hatching. A split of the dorsal keel of the ootheca releases the nymphs. At the time of hatching the cockroaches are white in colour and bear dark eyes, soon the integument becomes dark. The young ones are devoid of wings but in other respect they resemble the adult. These are called *nymphs*. It gradually grows into an adult. During this period the nymph sheds its integument at interval. This process of casting the integument is called *ecdysis* or *moulting* .It is brought about by a hormone secreted by prothoracic gland. The skin cast off by nymph is called *exuviae*. The interval between two moultings is called *stadium*. The form of nymph between two moults is called *instar*. There are about six or seven moults and an equal number of instars in cockroach. The metamorphosis is called paurometabolic development or gradual metamorphosis.

6

MUSCA

The houseflies of the genus *Musca* are very common in human dwellings. They are in plenty during the rainy season, though they appear early in spring and continue till late in winter. In winter most of them die, but many survive in warm places. There are several species of housefly, *Musca domestica* is common in Europpe, *M. nebulo* in tropical countries like India, Pakistan, Ceylon, Berma etc., *M. vicina* is common in all Oriental countries.

SYSTEMATIC POSITION

Phylum	-	Arthropoda
Class	-	Insecta
Subclass	-	Pterygota
Division	-	Endopterygota
Order	-	Diptera
Genus	-	*Musca*

Habits and habitat

The houseflies has a world wide distribution. They are abundently around human habitation and filthy and dirty places.

The houseflies are very active and abundant during spring, summer and autumn. They like light, and rest at night on the ceilings, walls and fixtures in the buildings. They become sluggish and decrease in number in November. The survivors pass winter in warm places. The houseflies can comfortably climb a vertical surface and walk upside down on a ceiling with their sticky feet. They are good fliers.

With the help of sponging type of mouthparts they take liquid food. They are unisexual. Their life-cycle show complete metamorphosis.

Morophology

Shape, Size and Colour. It looks somewhat ovoid without wings and nearly triangular with wings at rest. It is about 1.92 - 2.2 mm in width and 6-7 mm in length. Its dark grey body, with yellowish tinge on the ventral side, has four longitudinal lines on the throax and one black streak on the abdomen.

Division of body. Body is bilateral and is divisible into three regions:

1. Head.
2. Thorax
3. Abdomen

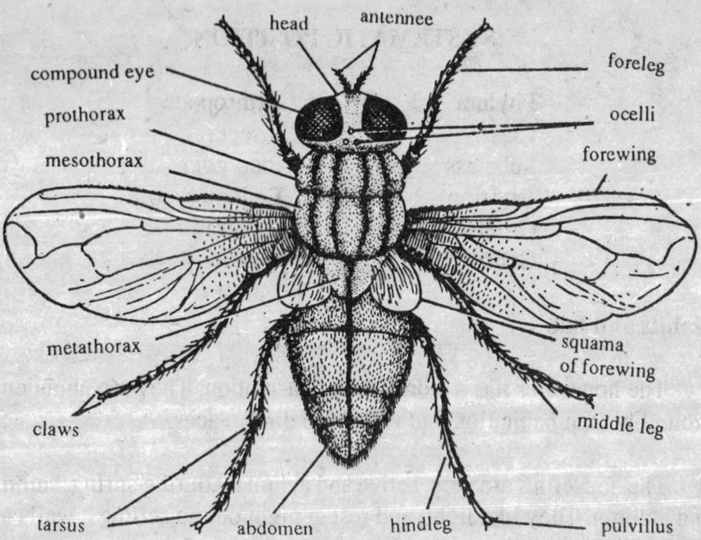

Fig. 6.1 *Musca nebulo*

1. *Head.* The head is as wide as rest of the body. It is mobile and covered with hair and bristles. It bears following structures.

(i) *Compound eyes.* The compound eyes are large occupying almost the whole of hemispherical head, thus provides a wide field of vision. Each eye bears 4000 ommatidia.

(ii) *Ocelli.* The ocelli or simple eyes are 3 in number. They lie on the top of the head (vertex) medially between the compound eyes.

(iii) *Antennae.* The antennae are located in a depression in the middle of the front side of the head.

Each antenna is three-jointed; the proximal joint is called *scape,* the middle *pedicel* and the distal *flagellum.* The flagellum is the largest and bears a three-jointed process called *arista.* The two basal segments of the arista are extremely small. The long distal segment bears two series of long hairs or *spinulae.* The antennae serve as sense organs.

(iv) *Mouth parts.* The mouth parts are sponging type adapted to suck a liquid food. It comprises the proboscis which is fleshy and retractile.

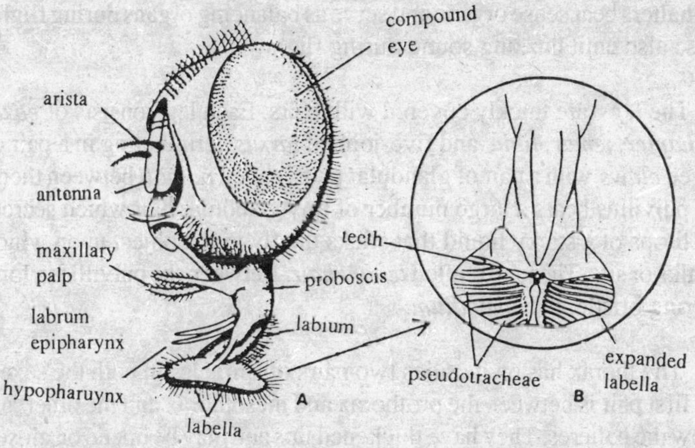

Fig. 6.2 *Musca* A= Mouth parts, B= Probascis enlarged

The proboscis consists of three parts. The basal *rostrum* middle *haustellum* and the distal *oral disc* or *sucker*. There is a movable joint between, the rostrum and haustellum so that the haustellum can be folded back beneath the rostrum in resting condition. The rostrum bears a pair of dark brown or black, club-shaped *maxillary palps* on its front side. The haustellum has a groove on its front side which contains the *labrum - epipharynx* and the *hypopharynx*. The edges of labrum - epipharynx are rolled to form a tube like *food channel*. The hypopharynx encloses the *salivary duct*. The oral disc comprises oval *oral lobe* or *labella*. The oral lobes are traversed by a network of fine grooves or channels, called *pseudotracheae*. All pseudotracheae lead to the centrally located *mouth aperture*, to which the tip of the food channel can be applied so that the liquid food collected by the pseudotracheae can be sucked up by the pumping action of the pharynx. The oral lobes are provided on the undersurface with minute, teeth-like structure for breaking up solid food.

For feeding, the proboscis is extended and its labella applied to the fluid food. The latter is drawn through the pores of the pseudotracheae into the food channel, whence it is sucked into the alimentary canal by the muscular pharynx.

2.*Throax*. The thorax comprises indistinct three segments *prothorax*, *mesthorax* and *metathorax*. It bears three pairs of legs ventrally and two membranous wings dorso-laterally. The wings are present on mesothorax and a pair of knowb-like structure called the *halters*. The halters bear sense organs and serve as balancing organs during flight. These also emit buzzing sound during flight.

The *legs* are thickly covered with hairs. Each leg consists of *coxa*, *trochanter*, *femur*, *tibia* and five-jointed *tarsus*, terminating in a pair of curved *claws* with a pair of glandular pads, called *pulvilli* between them. The pulvillus bears a large number of tiny glandular hair which secrete tiny drops of a sticky liquid that sticks the fly to the object upon which it walks or sits. These are called *tenant hair*. Between the pulvilli is a long tapering bristle, the *empodium*.

The thorax has on its sides two pairs of spiracles in both the sexes. The first pair is between the prothorax and mesothorax and the other just below the halteres. They have thickened lips and may be opend or closed to a limited extent by the action of special muscles.

3.Abdomen. There are ten segments in the abdomen, but the first has atrophied and the second is reduced, the third to the sixth are well developed and visible, but segments seven to ten are reduced and lie telescoped inside the anterior segments. In the visible segments the terga are large and extend ventrally. A pair of spiracles are present in the ventral edges of the terga on segments two to six. In the female the hidden segments seven to ten form a tubular ovipositor which protrudes and can be seen when the fly is depositing eggs.

Body Cavity

The body cavity is greatly reduced by air-sacs, thoracic muscles and fat bodies. It is represented by narrow spaces, the *sinuses*, among the viscera. It is not lines by epithelium and contains blood. It is, therefore, called the haemocoel. It is incompletely divided by a horizontal partition, the *pericardium*, into two compartments; the small dorsal sinus that surrounds the heart, and the large ventral sinus that contains the remaining viscera. The two communicate through a passage bounded by pericardial cells.

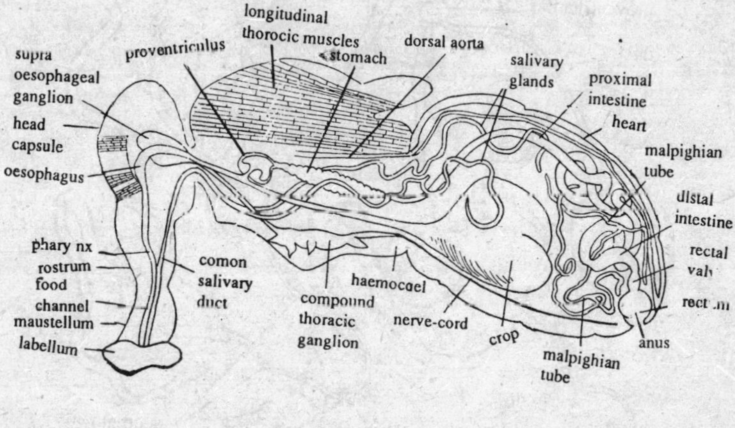

Fig. 6.3 *Musca* General anetomy

Digestive System

The digestive system comprises the alimentary canal and digestive glands.

1. *Alimentary Canal.* The alimentary canal is complete and comprising three regions: foregut, midgut and hindgut. The foregut and hindgut are ectodermal in origin and internally lined with cuticle

Mouth. It is situated in the middle of oral lobes, leads by way of a grooved channel into a pharyngeal bulb.

Pharynx. It is a tubular organ provided with muscles for sucking food. Just within the head capsule, it marges oesophagus.

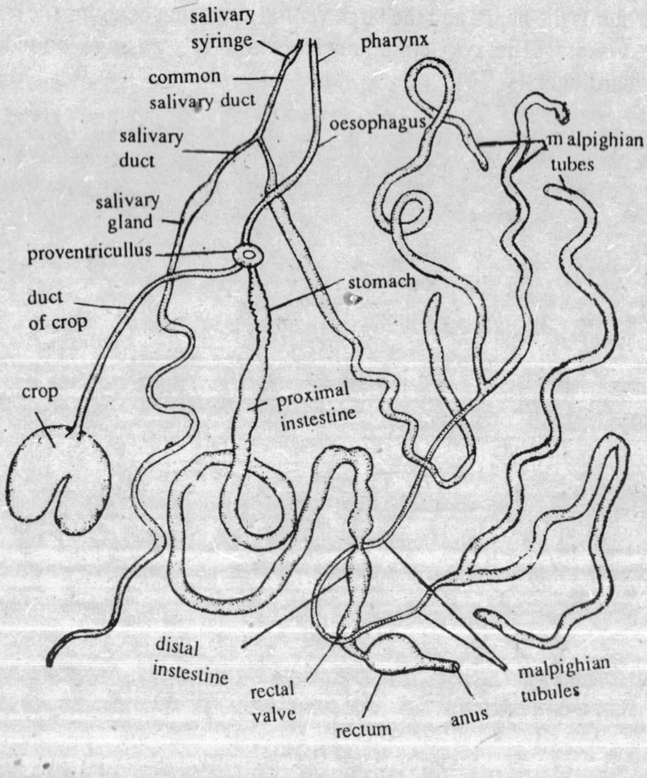

Fig. 6.4. *Musca*. Alimentary canal.

Oesophagus. It is also a tubular structure passing through the cerebral ganglia. After entering into thorax the oesophagus leads into the *gizzard* or *proventriculus*. Before opening into gizzard, the oesophagus gives off a straight, slender duct which extends the length of thorax into the abdomen, where it expands into a large bilobed muscular *crop*. *Pro*ventriculus. It is a small, circular chitinized sac lying in the prothorax, receiving oesophgus on its ventral side.

(ii) *Midgut*. It includes *ventriculus and proximal* intestine.

Ventriculus or chyli stomach. The proventriculus opens into the ventriculus. It is a straight, spindle-like sac, lines with columanr epithelium. The epithelium is folded to form a large number of *sacculi*. The ventriculus is followed by a long, coiled tube known as proximal intestine.

(iii) *Hind gut*. It comprises a *distal intestine* and *rectum*. The *distal intestine* is similar as proximal intestine but it is much shorter. It bears a rectal valve. The *rectum* is a dilated sac. It has tuberclated cuticle in anterior part, 4 rectal glands in middle part and thick muscular wall in terminal part. It opens out through the anus.

11 *Digestive glands*. There are a pair of *salivary glands*. These are wavy tube of uniform diameter. These are lined with cuboidal epithelium. Each gland gives off a *salivary duct* anteriosly. The two ducts unite to form a *common salivary duct*. The duct passes beneath the oesophagus. In the rostrum the common salivary duct is expanded to form the *salivary syringe that pumps saliva onward to the hypopharynx*.

Feeding and Digestion

During feeding the proboscis is extended and the two expanded oral lobes of the labella are placed on the exposed liquid food. If the food material is solid the fly first pours a little saliva or regurgitates droplets of liquid from its crop to liquefy it and then sucks the liquid. The liquid food fills the tubular pseudotracheae by capillary action. The pseudotracheae then converge into mouth, whence the food rises up the food channel due to the suction force set up in the muscular pharynx. The alimentary canal is long and coiled with the pharynx, oesophagus, crop gizzard, stomach or midgut and rectum. The two small salivary glands pour saliva on food to hydrolyse starch. Digestion and absorption mainly take place in the mid-gut.

Respiratory System

The respiratory system is hightly developed. It consists of three sets: the *spiracles, air-sacs and tracheae.*

Spiracles. These are 7 pairs of spiracles in a female and 9 pairs in the male. Of which first two pairs are thoracic and rest are abdominal. Each spiracle leads into a shallow *vestibule* which is partly separated by a valve-like mechanism from the *atrium* inside.

Tracheae. In the abdomen, the atria lead into the tracheae where as in the thorax the atria open into large thin walled *air sacs* from which small tracheae extends to muscles and other tissues. The thoracic air sacs send a number of *cephalic air sacs* in the head and a pair of very large *abdominal air sacs* into the abdomen. The cephalic and few thoracic air sacs remain expanded where as the others are collapsable. These are the reservoirs of air. The tracheae branch and rebranch to form the fine tubules, the *trachioles.*

The trachioles participate in the exchange of gases.

Circulatory System

The circulatory system of housefly is very simple and open type as found in other insects. There is a muscular tube or *heart* lying in a cavity i.e. the *pericardial sinus* along the mid- dorsal line. It extends the length of the abdomen and is divided into 4 chambers. Each chamber having a pair of opening through which the blood is sucked form the pericardial sinus. The anterior end is continued into a slender *dorsal aorta* or *anterior aorta.* The heart is supported by *alary muscles.* Besides heart there are 4 accessory pulsating organs called *auxillary hearts* at the base of each wings and a few in the thorax which assist in circulation of blood.

The dorsal aorta continue along the stomach and appears to terminate in a mass of cells behind the proventriculus. It pours the blood into the anterior part of haemocoel. The haemocoel is incompletely divided into a large *ventral sinus* and a small *dorsal sinus* or *pericardial sinus*, by a horizontal partition, the *pericardium.*

The blood is colourless and is called the haemolymph. It contains a fluid plasma and numerous blood corpuscles, the *haemocytes*. It does not contain any respiratory pigment. It helps in physiological exchange of materials.

Excretory System

In housefly the excretory system consists of one pair of yellowish-green tube-like structures known as *Malpighian tubules*. These are present at the junction of mid-gut and hind-gut. Each tubule is divided into two branches. The branches are very long, highly convoluted and can-not be separated from the fat-bodies.

Nervous System

In housefly the *supraoesophegeal ganglion, sub-oesophageal ganglion* and *circumoesophageal connectives* are fused to form a common mass which is perforated by an aperture for the passage of oesophagus. The ventral nerve cord bears a *compound thoracic ganglion* which is formed by the fusion three thoracic ganglia.

From brain a median *ocellar nerve* is originated and is supplied to ocellus. A pair of *antennary* paired *pharyngeal nerves*, paired *labial nerves* innervate their respective organs. Paired *cervical nerves* arise from the nerve cord before joining to the compound thoracic ganglion. It supplies to the muscles of neck. From compound thoracic gaglion a pair of each *prothoracic, mesothroacic* and *metathoracic dorsal nerve* arise which innervate the thoracic muslces. A pair of each of *accessory mesothoracic* and *metathoracic dorsal nerves* arise from the thoracic ganglion and suppliment the corresponding dorsal nerves in supply. To the legs, a pair of each *prothoracic, mesothoracic* and *metathoracic crural* nerves are supplied. Two pairs of *abdominal nerves* are given by the dorsal nerve cord behind thoracic ganglion in the region of thorax. In the abdomen a series of *abdominal nerves* are given alternately on right and left to supply abdomen. The dorsal nerve cord ends in the genital organs

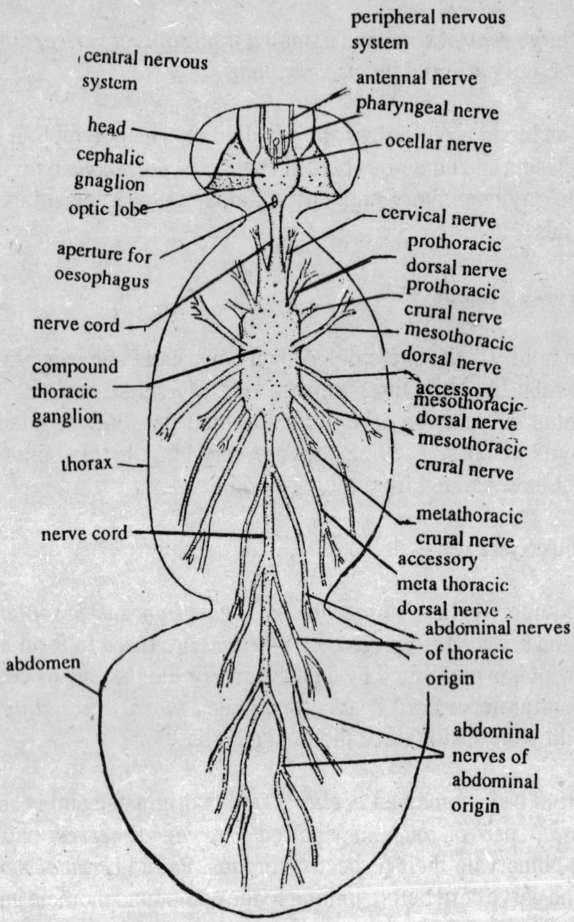

Fig. 6.5. *Musca*. Nervous System.

Reproductive System

The sexes are separate and are slightly different in size, the female being slightly larger than the male.

I *Male reproductive organs*. The male reproductive organs consist of a pair of testes, vasa deferentia, ejaculatory sac, ejacultory duct and penis.

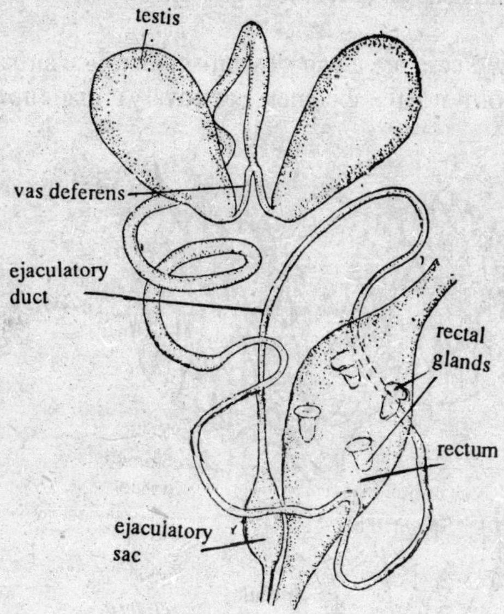

Fig. 6.6 *Musca*. Male reproductive organs.

(a) *Testes*. A pair of brownish, somewhat pear shaped or physiform testes are present in 5th abdominal segment.

(b) *Vasa deferentia*. From each testies a duct arises known as vas defesens and extends forward. The two vasa deferentia units to form a median tube, the ejaculatory duct.

(c) *Ejaculatory duct.* It is a much coiled structure, anteriorly it is wider. It enlarges posteriorly to form ejaculatory sac.

(d) *Ejaculatory sac.* It is a muscular sac like structure supported by ejaculatory sclerites. It helps in ejaculation of seman.

(e) *Penis* or *Aedagus.* The sclerites of 6-10 abdominal segments are modified into the external genitalia. There is a complicated penis. It is made up of a hollow *theca* followed by a delicate, hyaline *glans* bearing a male genital pore.

II *Female reproductive organs.* It consists of a pair of ovaries, oviduct, spermathecae and accessory glands.

(a) *Ovaries.* These are a pair of whitish, disc-like structure occupy major portion of the abdomen. Each ovary is made up of *ovarioles.*

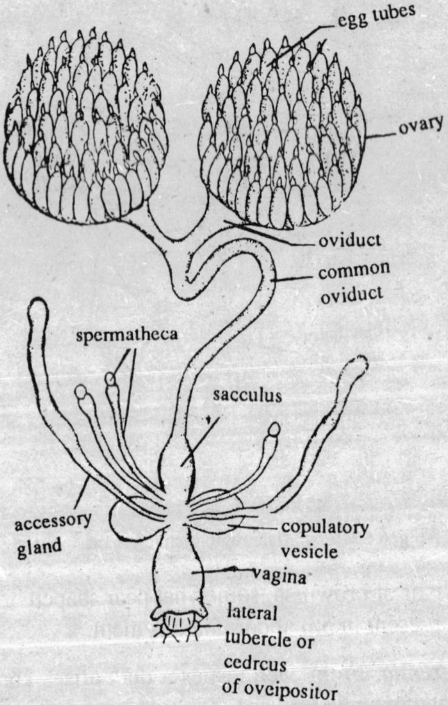

Fig. 6.7. *Musca.* Female reproductive organs.

and their number is 70 per ovary. Each ovariole consists of ova in various stage of development.

(b) *Oviducts*. All the ovarioles of each ovary lead into a very short, thin walled tube, the *oviduct*. The two oviducts units to form a curved *common ovdiuct*. It dilates to form *sacculus* which is followed by a still larger *vagina*. The vagina opens out through female genital pore present in the 9th abodominal segment.

(c) *Spermathecae*. These are three (one right and two left) long tubular structures which open into the genital tract at the junction of sacculus and vagina.

(d) *Accessory glands*. A pair of long, tubular glands lying near spermathecae. They secrete a sticky fluid that cover the eggs to adhere them together and to the substratum.

(e) *Copulatory Vesicles*. These are a pair of sac-like structure.

(f) *Ovipositor*. It consist of the abdominal segments 6-9 with long intersegmental membrane between. It normally remains with drawn into the 5th segment.

LIFE HISTORY

Copulation

The flies breed from June to October when they become very active. The greatest breeding activity occurs in the warmest months of August and September. The pairing of the male and female takes placc on the ground when at rest, and lasts for a few minutes. The male alights on the back of the female. The female inserts its ovipositor into the genital atrium of the male to receive the spermatozoa. The eggs are laid 6-8 days after copulation.

Oviposition. The rate at which the houseflies breed is amazing. A single fly lays 100-150 eggs. While each fly is able to laye such 5 or 6 deposits in one season. The female extends her ovipositor into the manure dung, decomosing fruits, vegetables etc. Whatever substance it selects and lays eggs.

Metamorphosis.

It is complete shows four stages of development as follows.

Eggs. The eggs are small, nearly white and elongate with one end slightly broader than the other. Each egg is about 1 mm. in length and bears two dorsal, rib-like, curved and longitudinal thickenings.

In warm weather the egg takes 8 - 24 hours to hatch.

Larva. The larva, specially called the *maggot,* emerges from a slit along one side of the egg, while growing up the larva cast off its skin twice, thus dividing the larval period into three *instars* (i) from hatching to the first moult that takes 20-30 hours (ii) between the first and second moult taking 24 hours (iii) from second mould till pupation taking about 3 days. The time between two moults is called *stadium*. The larva is a soft, legless, worm-like creature. It is conical is shape, being narrow in front and broad behind. The newly hatched maggot is 2 mm. long. Its body consist of 12 segments and an indisinct head. The head can be withdrawn under the first segment or pseudocephalon. It bears a pair of *oral lobe* bearing minute papillae-like *antennae*, a pair *maxillae* with photo-receptive papillae, a pair of hook-like *mandibles* and a small *labium*. It is *metapneustic* having only one pair of D-shaped posterior abodominal spiracles with two slit-like apertures in each. The anus is located ventrally on the last segment.

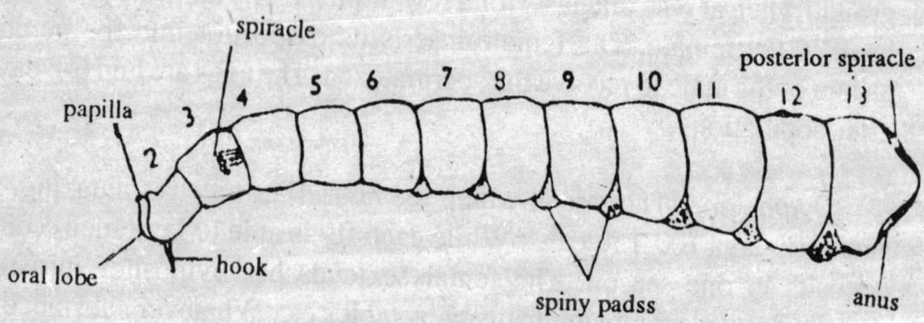

Fig. 6.8. *Musca*. Larva

On the anterior ventral border of each of the 6th-12th trunk segments, there is a semilunar pad best with minute. stuff spines. These *spiniferous pads* are locomotory organs and help the larva in moving backwards and forwards. The body of the larva is covered externally by cuticle.

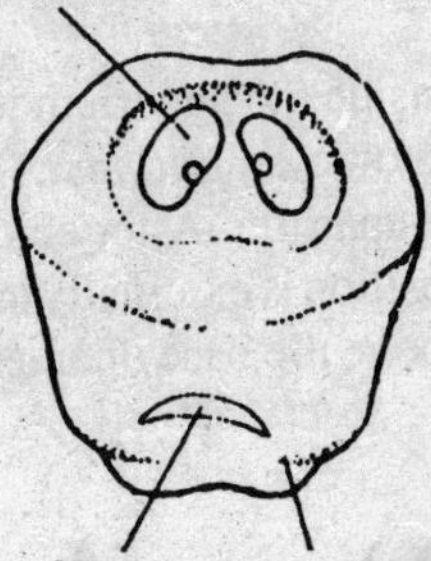

Fig. 6.9. *Musca*. Larval posterior and showing spiracle.

After first moult a second instar is formed. It is larger than first instar. Besides the posterior pair of spiracles which becomes larger, the 2nd instar acquire an anterior pair of spiracles also. Thus the larva is *amphipneustic*. After second moult, third instar is formed. The third instar is a full grown larva measuring 12 mm in length. The third instar grows in size and ready to pupate within 3 days.

Pupa. When fully matured the larva usually crawls away from its original moist place and travels to some dry and sheltered cervicc. Here it rests for sometimes. The emergence of pupa is proceeded by a general contraction of body. Some of the anterior segments are even withdrwan within the body. Thus the larva changes into a pupa without moulting, the last larval skin nardens to form an outer covering of *paparium* which encloses th pupa. Such a pupa is called *coarctate*. It has no chitinous covering of its own but only a soft pupal skin, the outer puparium has been formed by the last larval skin. The puparium is barrel-shaped and its becomes dark brown, externally it is segmented and shows traces of larval spiracles and spiny pads which become non-functional. The pupa takes in air by means of a pair of spine-like *pupal spiracles* projecting between the fifth and sixth segments of the puparium. The pupa is absolutely immobile, and the pupal stage last from 4 to 5 days. During this time internal changes take place, the larval organs are broken down or *histolysis* occurs. The *imaginal buds* of the larva begin to form the organs of the adult, the *histogenesis* occurs in the pupa. Imaginal buds are dormant cells, they are stimulated by a hormone of prothoracic endocrine glands wich become active only during metamorphosis and make the imaginal buds grow. By these process the adult fly or *imago* is formed in the pupa.

When the fly is ready inside the puparium it has a sac like structure on the front of the head known as *ptilinum*. This process aid in the emergence of imago from puparium. After emergence, the ptilinum is withdrawn in the head. Its position is externally indicated by a crescentric slit called *lunule* situated above the base of antennae in the form of 'U'

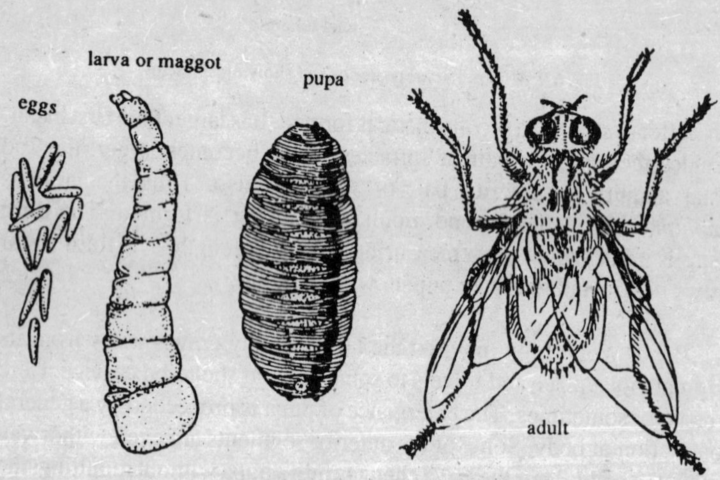

Fig. 6.10. *Musca*. Life history

Imago. It is somewhat smaller than the adult. The wings are less developed and the colour is light. In the contact of air the wings become hard and somewhat large. The body increase in size and within one or two weeks it becomes adult fly and ready for egg laying.

Economic Importance.

The housefly is a very harmful insect besides causing a lot of annoyance, it carries microganisms of several human diseases as it feeds on all eatables as well as manure, faecal matters and decomposing matter. A single fly carries about 33 million germs in its intestine and more that 100 million germs on its legs etc. 1. *Dissemination of germs.* They caog spread various diseases by carrying them of on the hairy surface of its body, its body legs etc. The germs are carried in droppings, vornit spots or via saliva. The diseases are tuberculosis, diarrhoea, typhoid, chlorea, leprosy , plague, anthrax, dysentery, ganoorrhoea etc.

Larvae of housefly ingest eggs of a nematode, *Habronema*, and the infection is passed through pupae to adults which transmit it to the horses and to the eyes of children causing conjunctivities.

Sometimes the eggs are laid on open wounds, where they hatch out as larvae when infected fruits or vegetables are taken, larvae hatch out in the gut, perforate the mucous lining and may cause haemorrhage. This attack by larvae on human is called *myiasis*

2. *Contamination of food.* The housefly contaminates our food material by transmitting the infective stages of *Taenia, Ascaris* etc.

Control of house flies

The houseflies are the dangerous enemies of human beings. Therefore, it is very essential that the growth and increase of houseflies and the infection by them must be stopped. It can be done in the following ways:

(1) *Preventing their breeding.* Elemination of the breeding media is the first and most important step in fly control. This can be done by:-

(a) *Physical method.* All the filth, where eggs can be laid, should be kept covered and periodically removed every week. Open toilets should be discouraged.

(b) *Chemical method.* The manure or rubbish it accumulated for a longer time, it should be treated chemically with lime, copper-sulphate, crude oil etc. to prevent the breeding of flies.

2. Destruction of adult flies

(a) *Mechanical methods.* Flies can be collected by baits and then killed. Fly papers, and fly swatters can also be used.

(b) *Chemical methods.* Spraying of houses, barns etc. with DDT, benzene hexachloride etc. is useful. A few drops of 3% formaldehyde solution in a plate of milk attract and kills flies. Repellants like creosote oil and borax, keeps the flies away. A mixture of coaltar and carbon disulphide can be applied to the wounds of animals. Fumigation by neem leaves is also useful.

(3) *Protection from infection by flies.* All types of filth should be kept away or duly covered. The houses must be properly screened and food, fruits and vegetables must not be exposed in houses or markets.

7

MOSQUITOES

Everyone is familier with mosquitoes. These are common in tropics and found in marshes and damp places. They are a nuisance in places where they occur in large number because of their biting habits. There are 29 gnera and 2000 species of mosquitoes in world. The important genera, as for as human health is concerned, are *Culex, Aedes, and Anopheles.*

SYSTEMATIC POSITION

Phylum	-	Arthropoda
Class	-	Insecta
Subclass	-	Pterygota
Division	-	Endopterygota
Order	-	Diptera.

Habits and Habitat.

Mosquitoes are found in damp and marshy places. They are abundant in tropics and sub-tropics. They are active in summer. Mosquetoes are nocturnal, as they come out in night for feeding. During day time they hide in hollows of trees, bushes, dark places of buildings etc. The female feed on the blood of man and other large animals where as male suck the juices of flowers and fruits. They are oviparous and lay eggs in water. Most of the mosquito die in winter and extreme hot months of May and June. Some survive hiding in protected places. The mosquitoes can fly for 2-3 hours continuously and cover 20-30 km.

Morphology

Shape, Size and Colouration. The mosquiot has a slender, delicate and long legged body. It measures 3-4 mm in length. The body colour is greyish black and the wings ogf *Anopheles* carry dark spots, *Aedes* has no spots an wings but the body has a black and white stripes.

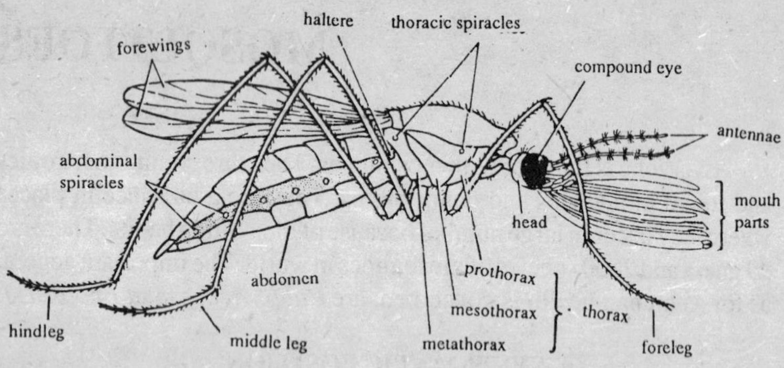

7.1. *Culex.* Female

Division of body. Like other insects, the body is divisible into three parts.

1. Head
2. Thorax
3. Abdomen.

1.*Head.* The head is globular and highly mobile on a slender neck. It bears a pair of compound eyes, antennae and mouth parts. (i) *The compound eyes* are large and kidney shaped. They occupy a considerable part of the head. The ocelli are absent. The top of the head has an epicranium below which is clypeus. (ii) *The antennae* are filiform, each with 15 joints. The basal segment is the scape which is concealed by a very large globular second segment, the pedicll containing a *Johnson's organ* which is auditory in function, the other 13 joints form a *flagellum* having many bristles lying in rings. The bristles are longer and much more numerous on the antennae of males giving them a bushy appearance. The antennae of female are with only a few, short bristles. (iii) *Mouthparts*. The mouth parts are of *piercing* and *sucking type*. contained in a proboscis. These are:

(a) A pair of mandibles

(b) A pair of maxillae

(c) Labium

(d) Labrum and epiparynx

(e) Hypopharynx

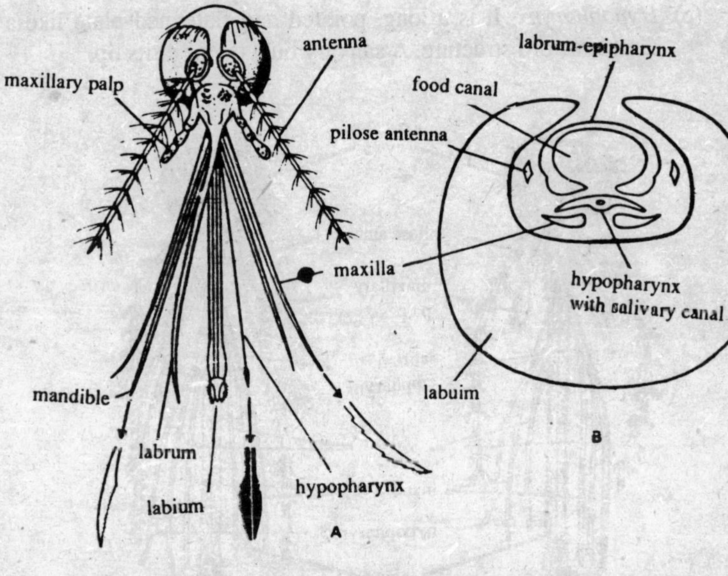

7.2. Mouth parts of Mosquito. A = Frontal view, B = T.S. of same

(a) *Mandibles*. The paired mandibles are long, thin and needle-like styles with fine blade-like distal edges. These act as piercing organs and puncture the host skin. The mandibles are, therefore, found only in females and are absent in males.

(b) *Maxillae.* These are also elongated needles very much like the mandible but ending into saw-like blades. These bear a pair of 5 segmented maxillary palps.

(c) *Labium.* This is formed by the fusion of 2nd pair of maxillae and is in the form of a median ventral, elongated and fleshy *proboscis sheath.* It is like an open gutter of half tube with a groove on its dorsal surface, The tube lodges the mouth-parts and is roofed over by labrum and epipharynx. The free distal end of the proboscis ends in a pair of white pointed lobes known as *labellae.* The labellae are tacile organs clothed with numerous sensory hair.

(d) *Labrum and epipharynx.* The labrum and epipharynx are fused together forming a stiff structure with pointed tip. It lies dorsally abover the groove of labium and roofing it. It bears ventrally a groove. The groove and hypopharynx forms a food channel.

(e) *Hypopharynx.* It is a long, pointed and flatenned plate like a double sword structure. A salivary duct opens at its tip.

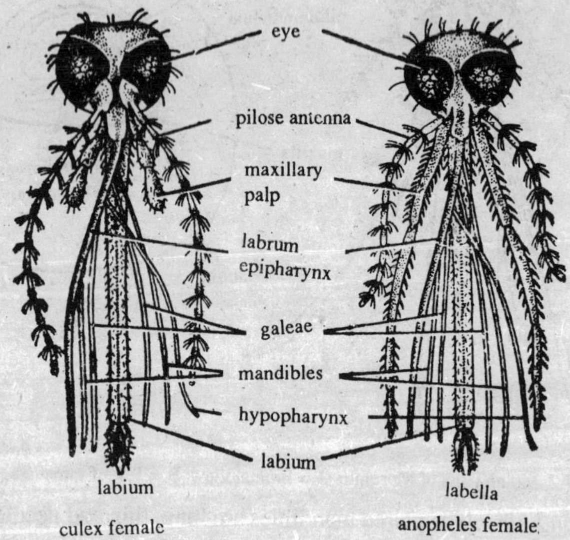

7.3. Head and mouth parts of female *Culex* and *Anopheles*.

2. *Thorax.* The thorax is as usual made up of three segments. The prothorax and metathorax are small whereas the mesothorax is very large

and its tergum consists of three sclerites: *scutum, scutellum* and *post scutellum*. Each segment bears a pair of legs. The legs are long, slender and delicate. They have short coxae, long tarsi and simple claws.

One pair of wings are present on dorsolaterasl side of mesothorax. These are membranous. The mesothorax bears a pair of *halters* or *balancers*. These represent rudimentary hind-wings, useless for flight and probably sensory in function. These are supposed to produce characteristic sound.

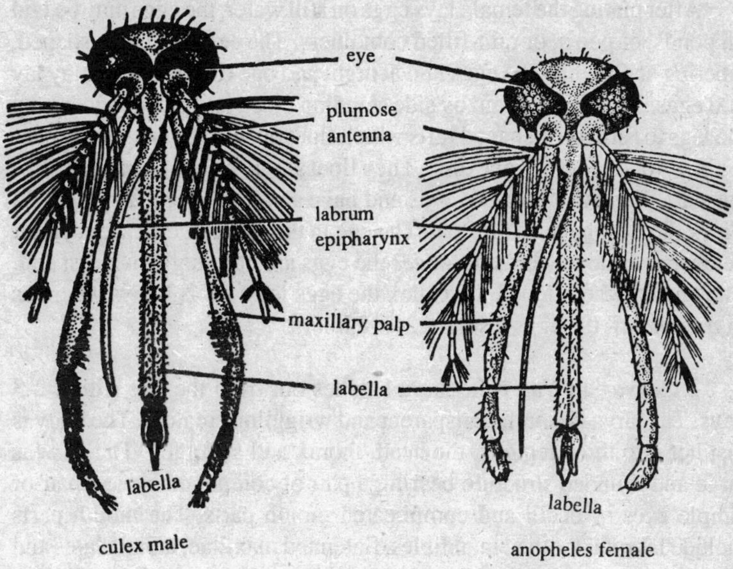

7.4. Head and mouth parts of male *Culex* and *Female*

3.*Abdomen*. The abdomen is slender and made up of 10 segments, the first one is fused with metathorax so 9 segments are visible. In male last 3 segments are modified by genitalia. It bears an *anus* on the 8th segment and a *genital pore* on the 9th segment.

Life History

The life history in mosquitoes coprises four stages:

1. Egg
2. Larva
3. Pupa
4. Adult

The female always takes a blood meal before egg-laying to have extra nourishment essential for this purpose.

1. Eggs

After mating the female lays eggs on still water, the eggs may be laid on ponds, or pools, or rain-filled containers. The eggs are cigar-shaped, tapering at one end. The eggs laid at night and one *Culex* female may lay 300 eggs. The eggs are laid by side standing erect, and glued together by the legs to form bost-shaped rafts which float on water. Female Anopheles lays 40-100 eggs at one time. They float singly do not form a raft. The eggs are pointed at both the ends and have a pair of lateral *air floats*. In *Culex* the eggs have no air float. The egg in the raft are vertical and, their heavier ends downwards. In *Aedes* the eggs are laid singly in moist soil. They hatch when floded. In Culex the eggs least for 2-3 days where as in *Anopheles* the eggs lasts for 24-48 hours.

2. *Larva.* In *Culex* the larva hatches out from the egg within 2-3 days. The larva is small, transparent and wriggling creature. The body is distinct into three regions, the head, thorax and abdomen. The *head* is large and rounded structure bearing a pair of compound eyes, a pair of simple eyes or ocelli and complicated mouth parts. The mouth parts include labrum, toothed mandibles, flatenned maxillae, labial plates and two dense mouth brushed. The larva, thus, subsists on solid food. The thorax is unsegmented but bears three lateral pairs of unjointed tubercles, each bearing a tuft of bristles. The abdomen comprises nine segments, each bearing a tuft of bristles. The 8th segment bears a long, tubular, chitinous respiratory siphon The siphon bears a *comb* of very small scales at the base 2 large hair tuft and two tows of flatenned spines called *pecten*. At the tip of the siphon there are two *spiracles* leading into the tracheae. Around the spiracles are five leaf-like lobes which can close over the spiracles to prevent water from entering. The respiratory system is *metapreustic* in which only the last pair of abdominal spiracle is open. At the end of the 9th segment is an anus surrounded by four leaf-like *tracheal gills*. The tracheal gills bear tracheae instead of blood vessels.

The 9th segment bears a tuft of *dorsal bristles* at its tip and ventrally a bushy tuft of bristle called *ventral brush.*

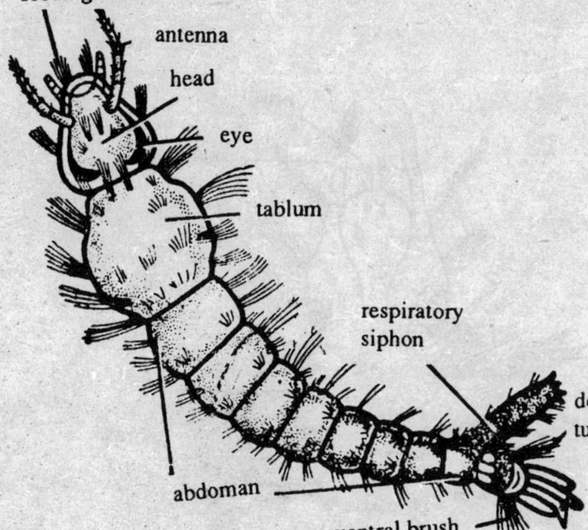

7.5. Larva of *Culex.*

The larva a *Anopheles* is without a respiratory siphon instead the 8th segment bears a raised chitinous quadrilateral plate with two spiracles. The spiracles are surrounded by five small leaf like flaps. Near the spiracles, there are small bristles which form a pair of *pecten. Comb* is absent. The larva has *palmate bristles or hairs* forming a pair of tufts on each thoracic and most of the abdominal segments. A palmate bristle consists of a short stalk from which a number of slender and flat leaflets tradiate. In *Anopheles* the larva can breathe and feed lying parallel with the surface of water sueeface as in *Culex* it floats in water obliquely with head lowermost.

During this period the larva moults for four times. A full grown larva is baout 12 mm long.

3. Pupa

After the last moult, the larva becomes pupa. This is the resting stage. The pupa has a large rounded anterior part consisting of the head, thorax and a slender, bent posterior abdomen. Thus it is comma-shaped. It floats in water with the cephalothorax uppermost and the abdomen curved

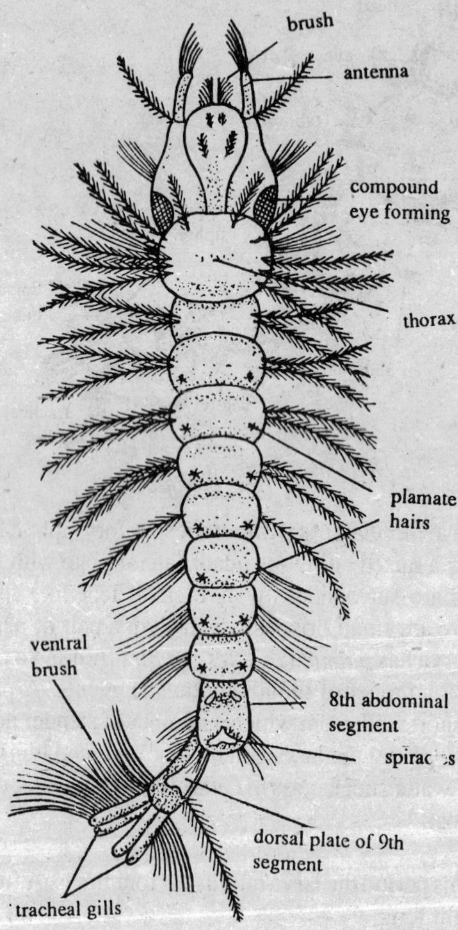

7.6. Larva of *Anopheles*.

beneath it. On the mid- dorsal side of the cephalothorax is a pair of tubular *respiratory trumpets* which are broader at the distal end. They communicate with an anterior pair of thoracic spiracles. It swims actively with a pair of large, chitinous and leaf-like *paddles* or *fins*. The pupa is unable to fedd. It flutters up and down in water when disturbed, hence, it is called a *tumbler*.

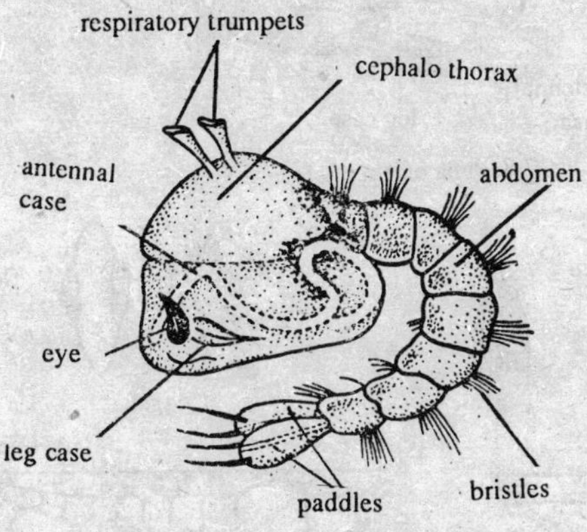

7.7. Pupa of *Culex*.

In the pupa of *Anopheles* the respiratory trumpets are short and broad and the abdomen is more strongly curved than that of pupa of *Culex*.

The pupal period lasts for 1 to 3 days. On this period the pupa is transformed into a complete adult. Most of the larval orgnas in the pupa arc broken down. The process of breaking down and disintegration of larval orgnas is called *histolysis*. The adult organs arc formed. For the formation of adult organs groups of formative cells are set aside in the larva, they are called *histoblasts* or *imaginal buds*. This process is called *histogenesis*. Through the transparent skin of the pupa, thc adult organs, such as compound eyes, antennae, wings and legs are visible.

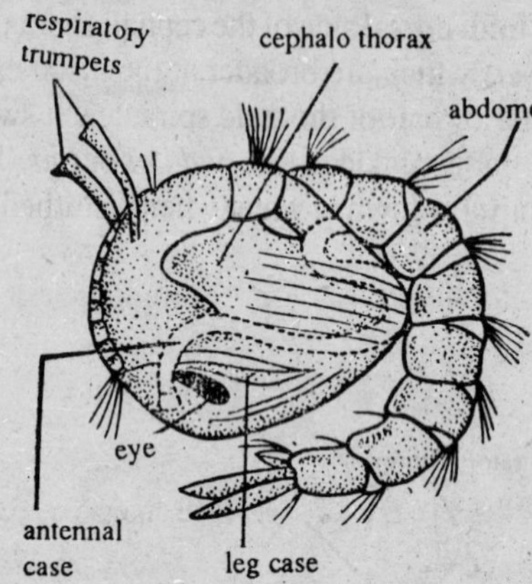

7.8. Pupa of *Anopheles*.

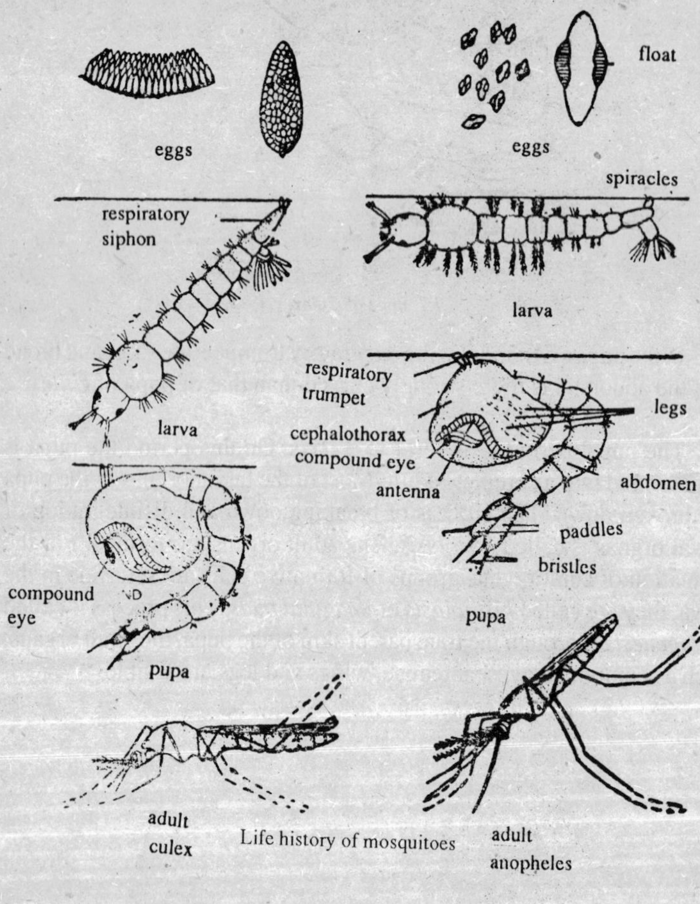

Life history of mosquitoes

4.Imago

The pupal stage lasts for a short duration, usually of two or three days. After the completion of metamorphosis, the pupa is transformed into the perfect adult insect, called *imago*. The pupal skin now splits along the back between the two respiratory trumpets. The imago emerges out with well-developed wings and after a few moments flies away.

The complete life cycle, from egg to adult, takes from ten to several days. The female lives for one month or more and the male for about a week.

Table showing differences between the life history of *Culex, Anopheles* and *Aedes*.

Culex	*Anopheles*	*Aedes*
I. Eggs		
1. Eggs are laid on the surface of water in a mass called the egg-raft.	1. Eggs are laid singly on the surface of water.	1. Eggs are laid singly in mosit soil and hatch when flooded
2. Eggs are cigar shaped and lack floats.	2. Eggs are boat shaped. Each has a pair of lateral float	2. Eggs are spindle shaped and lack floats.
3. Eggs lie vertically in the raft.	3. Eggs float horizontally on the surface of water.	3.
4. Number of eggs laid at a time varies from 200 to 300.	4. Number of eggs liad at a time varies from 40 to100.	4. Number of eggs liad at a time varies from 200 to 400.
II. Larva		
5. Larva floats in water obliquely with the head lowermost.	5. Larva floats horizontally on the surface of water.	5. Larva floats in water obliquely with head lowermost.
6. Without palmate hair	6. With palmate hairs on	6. Without palmate hair

	abdomen	
7. Larva bears a long, tubular, chitinous respiratory siphon on the 8th abdominal segment for breathing air. The siphon bears more than one pair of hair tufts.	7. Larva has, on the 8th abdominal segment, a rised chitinous plate bearing two spiracles for breathing air.	7. Larva has a respiratory siphon that is relatively short and stout and bears only one pair of hair tufts.
III.Pupa		
8. Pupa bears long respiratory siphons on the cephalothorax.	8. Pupa bears relatively short and broad respiratory siphons on the cephalothorax.	8. Pupa bears long respiratory siphons on the cephalothorax.
9. It is colourless.	9. It is greenish.	9. Pupa is colourless.
10. Each paddle bears a single long bristle.	10. Each paddle bears one long and one short bristle.	10. As in *Culex*
IV Imago.		
11. Greyish body.	11. Greysih bairy body.	11. Greyish black body.
12. Wings lack dark spots.	12. Wings have dark spots.	12. Wings are without spots.
13. Imago rests with the body parallel to the surface. Body hump backed when at rest.	13. Imago rests with the body making an-actue angle with the surface. Body is not hamp-backed when at rest.	13. Imago rests with the body parallel to the surface. Body is hump-backed when at rest.
14. Proboscis is bent downward, when imago is at rest.	14. Proboscis is in line with the body, when imago is at rest.	14. Proboscis is bent down when the imago is at rest.
15. Abdomen bears scales on the sterna.	15. Abdominal sterna lack scales.	15
16. Maxillary palps are short in the female and longer than proboscis in the male but not clubbed.	16. Maxillary palps are as long as proboscis in both sexes and clubbed in the male.	16. Maxillary palps are short in the female and longer than the proboscis in the male but not clubbed.

17. Scutellum is tri-lobes.	17. Scutellum is evently rounded.	17. Scutellum is tri-lobed.
18. Thorax is without markings.	18. Thorax lacks markings.	18. Thorax has silvery or white markings.
19. Tip of abdomen of female is blunt with cerci retracted.	19. Tip of abdomen of female is blunt with cerci retracted.	19. Tip of the abdomen of the female is poin-ed with cerci protruding.
20. Spreads filariasis and encephalitis.	20. Spreads malaria.	20. Spreads yellow fever and dengue.

Economic Importance

The mosquitoes are definit nuisence as they inflict painful and irritating bites besides sucking blood. Most of them act as transmitting agents or vectors for some diseases of man and other animals. The diseases caused by mosquitoes in man are malaria, yellow fever, filariasis, dengue, encephalities, leprosy etc.

(i) *Malaria*. Malaria a deadly human disease. It is caused by a protozoan, *Plasmodium*. It is transmitted by female *Anopheles*. In malaria there is high recurring fever accomplished by chills and shivering.

(ii) *Yellow fever*. Yellow fever is confined to Africa and south America. It is caused by a virus and the vector is *Aedes aegypti*. The symptoms include high fever, internal bleeding, severe headache, and pains in the bones, anaemia, severe jaundice and vomitting of blood and bile occur.

(iii) *Filariasis*. Filariasis is caused by a helminth parasite, *Wuchereria bancrofti*. Its vector is *Culex fatigans* The disease begins with a slow inflammatory swelling of limbs, scrotum or manmmae. Filariasis may eventually cause *Elephantiasis*, if left unattended.

(iv) *Dengue*. Also called back bone fever is caused by a virus and is carried by *Culex fatigans* and *Aedes aegyti*. The symptoms are sudden high fever with rash on the face and severe headache, pain in muscles and joints etc.

(v) *Encephalitis*. Encephalitis is caused by a virus and is spread by various species of *Culex* and *Aedes*. The symptoms are headache, high fever, drowziness and inflammation of brain.

(vi) *Leprosy*. According to the scientiests of JIPMER, leprosy can be spread by mosquitoes.

(vii) *Dermatobia*. Dermatobia is a disease of the skin of man and cattles of central and south America. *Psorophora* a, mosquito, carries the eggs of a bot-fly, *Dermatobia* (as it lays egg on the body of mosquito). When the mosquito botes, the eggs of bitfly quickly on the human skin and the larvae penetrate the skin casuing swelling and cutaneous myasis.

Control of Mosquitoes

Following general methods may be used for the control of mosquitoes:

1. *Elemination of breeding places*. The larvae and pupae develop in the water, therefore, elemination of breeding places is important. Small ponds, ditches etc. can be filled up with mud. Stagnant water,swampy areas, marshes must be drained of. Water should not be allowed to be stored in open cans, barrels, buckets etc. Bushes, shrubs etc. should be cleared.

2. *Destruction of larvae and pupae*. It is easier to kill the aquatic larvae and pupae than the adults. It can be done by spreading over water surface with cruide oil, kerosine oil, etc.. Oil emulsions of DDT, BHC etc. are effective larvicides. Biological methods can also be employed such as larvicidal fishes like *Gambusia, Guppy*, trouts, sticklebacks etc. can be introduced in the water resevior. Aquatic insectivorous plants, such as bladderwort (*Utricularia)* can also devour the larvae and pupae of mosquitoes.

3. *Destruction of adult mosquitoes*. This can be done by hands, traps fumigation, spraying etc. Fumigation with tarcamphor, cresol, sulphur or other derivatives of neptha burn to produce poisonous smoke which kill the adults. Spraying with DDT, gammexane, melathione etc. remain effective.

4. *Sterlization*. Sterlization of mosquitoes in Japan has been successfully done.

5. *Personnel protection*. It can be done by ; (i) The exposed parts of the body may be protected by the use of gloves, soxes, boots etc. (ii) Use mosquitoe repellents like odomos, mustard oil etc. (iii) Mosquito nets should be used, especially at night (iv) The house should be built on a high grounds with proper drainage.

8

APIS (HONEY BEE)

Apis is a social insect living in colonies, build by themselves. A colony may have more than 50,000 individuals. They are semi-tropical insects that have adapted themselves in temperate climates, where they survive cold weather by forming themselves into winter balls within the hive.

These are five well recognized types of bees found in the world.

(i) *Apis dorseta* (Rock bee)
(ii) *Apis florea* (Little bee)
(iii) *Apis indica* (India bee)
(iv) *Apis adamsoni* (African bee)
(v) *Apis mellifica* (Eurpoean bee)

Out of these, three arc common in India. They are *A.dorseta, A. florea* and *A.indica. A.dorseta* is the largest variety with average size of about 20 mm. It builds large comb on tree branches, high buildings etc. These are migratory. Species as during June and July they swarm to the hills, but in winter they come back. The worker build a fresh comb every time. A single comb may yield a fresh comb every time. A single comb may yield 60 pound of honey per year.

A.Florea is a minature of the rock bee. It is a plain species and builds small comb on the tree branches, in bushes etc. They yield hardly a few ownce of honey per comb per year.

A.inica is of commonest occurance on the plain and forests. It builds several parallel combs in protected places like hollow of trees, caves, in rocks etc. The average output of honey is 6-7 pounds per colony per year.

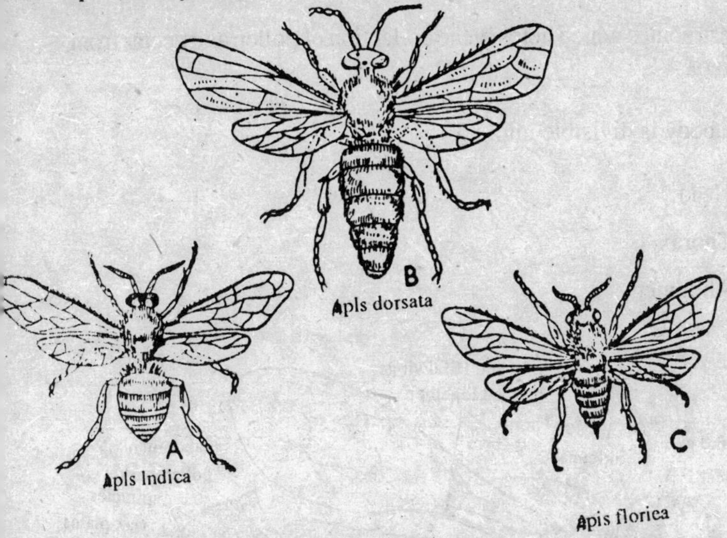

8.1. Workers of three species of honey bee.

SYSTEMATIC POSITION

Phylum	-	Arthropoda
Class	-	Insecta
Subclass	-	Pterygota
Division	-	Endopterygota
Order	-	Hymenoptera
Genus	-	*Apis*

Castes of honey bee.

The nest of honey bee is known as been-hive. A hive in summer consist of 32 - 50 thousand or more individuals. A colony may be *weak* or *strong* according to the number of worker bees. In a colony there are three types of individuals namely the queen, workers and drone. Due to the existence of several morphological forms, bees are said to be polymorhpic species.

Structure of worker bee.

The worker is the smallest member of the colony. Its body is densely covered with hairs. The colour is usually brown or black bearing several peculiar strucutes which helps in the collection of pollen and nectar from the flowers.

The body is divisible into:

1. Head.
2. Thorax.
3. Abdomen.

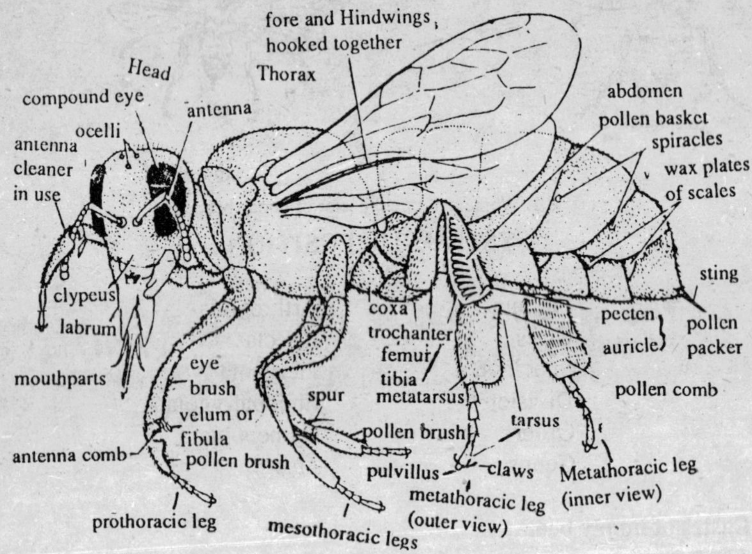

8.2 *Apis*. Worker bee in lateral view.

1. *Head*. The head is flatenned triangular widest crosswise through the upper corners, which are capped by the large compound eyes. The compound eyes are present on dorso lateral sides of the head. Each eye has several *ommatidia* (4900 in queen, 6300 in worker and 13000 on drones). Three ocelli are present in the middle of the face. The autennae are many - jointed and inserted in the middle of the head. They have many of factory pits that provide a keen sense of small *mouth parts*. The mouth parts attached to the lower part of the head and are of *chewing and lapping type*. These are modified for collecting the nectar and the pollen. Mouth parts consist of *labrum, epipharynx, mandibles, maxillae* and

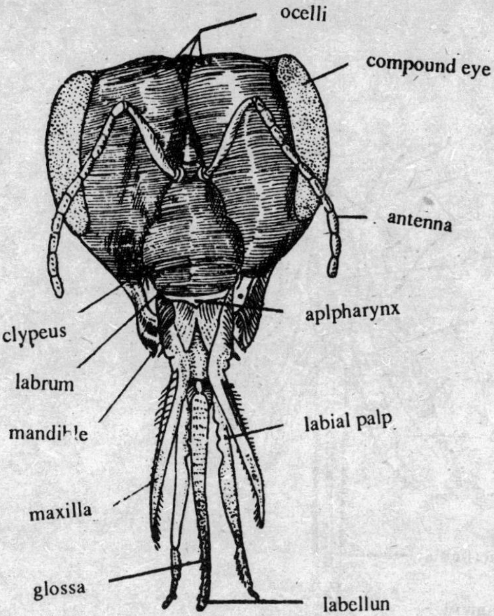

8.3 *Apis*. Head

labium. Labrum lies below the clypeus, below the labrum is fleshy epipharynx which is an organ of taste. At the sides of labrum are the two heavy *mandibles* which work sidewise. In the workers they are spoon-shaped as they lie, but they are sharp, pointed and toothed in the queen and drone. The function of mandibles are to mould wax and manipulating the pollen. In the *maxilla,* lacina is absent while maxillary palps are vestigial and the *glaea* is elongated and blade like. In the *labium,* the paraglossae are very much reduced and glossal are vary much elongated. they are united, hairy and form a *honey spoon* or*labellum* at the termination. The labial palps are well developed and they help to make the ligula up. The whole apparatus is protected the galeae of maxillae. When the bee sucks up nector, the terminal lobes of labium and maxillae are pressed close together as to make a tube between them. The food is then taken in by a sucking action of the pharynx. At the tip of the tongue these occurs lobella and these is a groove running along the entire length on the ventral side, within the tongue this groove extends into a double barrel tube. A flexible chitinous rod lies along the back wall of this channel which is itself provided with a still finer groove along its ventral surface.

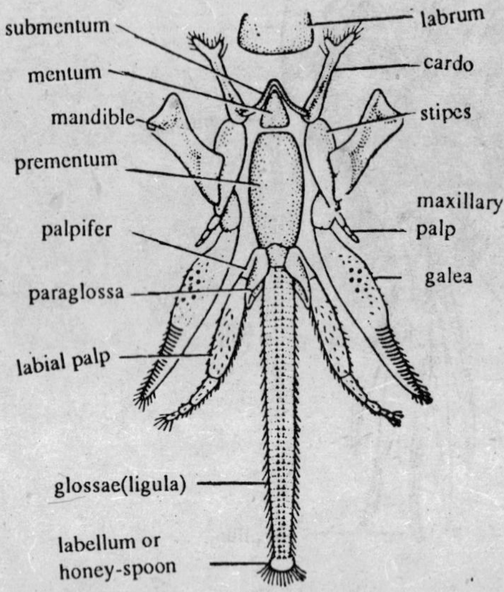

8.4. *Apis*. Mouth parts.

2. *Thorax*. The thorax is large and strong. It bears two pairs of wings and three pairs of legs.

The *wings* are small, narrow, membranous and transparent. They lie flat over the back at rest. They have a much modified and reduced venation. The fore and hind wings of each side are interlocked by hooks held in a groove, so as to work together during flight. Wings may vibrate over 400 times per second during flight. A worker is capable of long flight, even 12 km.

There are three pairs of *legs,* which are densely covered with hairs. The forelegs are called *prothoracic*, mid-legs as *mesothoracic* and hind legs are *metathoracic*. Each leg consists of five parts, *coxa, torchanter, femur, tibia* and *tarsus*. The tarsus is five jointed and terminating in to the ***claws*** and pulvillus.

(i) *Prothoracic legs*. *Tibia* of the fore-legs possesses distally at its anterior face a number of stiff bristles which forms *pollen brush*. From its posterior face there is also found a movable plate-like process or *velum*, which fits over a circular notch in the upper part of the part of the first tarsal segment. The *velum* and *antemna-comb* together serve as antennacleaners. On the anterior surface of the first tarsal segment there are also some bristles which are termed as *eye-brushes* and serve to remove particles and pollen from the outer surface of the compound eyes.

(ii) *Mesothoracic leg*. The midleg has all the segments as seen in the foreleg. The tibia bears a *brush* on its inner surface and a spine-like *pollen-spur* on its distal end. The spurs of both the midlegs are used to remove pollen from the *pollen* baskets of hindlegs and to disloge wax from wax pockets on the ventra surface of the abdomen.

(iii) *Matathoracic legs*. Each *metathoracic leg* has a large tibia with a cavity with bristles forming a *pollen basket* or *corbicula* used for storing pollen during collection, at the distal end the tibia has a row of stiff bristles called *pecten* below which is a flat plate, the *auricle*. The pecten and auricle form a *pollen packer* to convey pack pollen into the pollen basket.

3. *Abdomen*. The abdomen is oval and less hairy than the rest of the body. It bears six distinct segments. It possesses *wax glands* and *sting*.

(i) *Wax glands*. The glandular area secreting wax lies on the ventral surface of the last four visible segments of the abdomen. The wax is secreted through minute pores in the form of flat scales. The wax is masticated by the mandibles before its use for building the cells of the honeycomb.

(ii) *Sting or terebra*. In the worker bee, *ovipositor* changes so as to take the function of sting, which for injecting poison. The actual sting or *terbera* consists of a central poison-canal enclosed by a *stylet sheath* and two ventral *stylets* or lancelets. *Stylets* are grooved on dorsals side and form two ventral *guide sails*. The distal free termination of *stylet* and *sheath* are covered by *spines* or *barbs*. At the base, there is an expansion, which is called a bulb, while sting is continued into a pair of lateral basal arms. The sting is attached to a set of three plates :

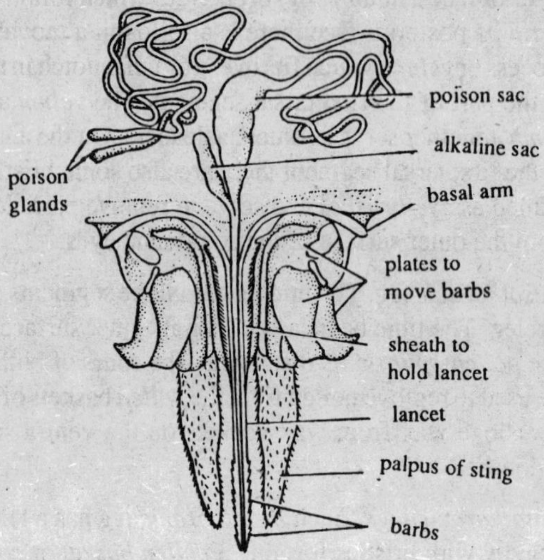

8.5. Apis. A sting.

(a) The inner most pair is that of *oblong plate* posterior in position and representing the divided 9th sternum. To each is attached the basal arm of the style sheath of its side. (b) Two *triangular* or *fulcral plates* represent the reduced sterum of 8th segment and to each is attached the corresponding arm of the stylet. (c) The large *quadrate plate* lies dorsally to the triangular plates and at the posterior angle. It represents the sternum of 9th segment these three pairs of plates act as the sternum of 9th segment and also function as levers. Two glands are also associated with the sting. A filiform poison gland pours its acidic secretion into a large poison-sac, which discharges its contents into the bulb. There is also an alkaline gland, which opens separately into the *sting bulb*.

Structure of Queen.

The queen is 15-20 mm long. It is similar to the worker except elongated, tapering abdomen, short legs, small wings, short mouth parts and an ovipositor which serves for egg laying. The reproductive organs are well developed. The waxglands and the pollen collecting apparatus are absent. The life-span of a queen is 2-5 years and during this span is lays 1.5 million eggs.

Structure of Drone

A drone is 15-17 mm long and stouter than worker. It has *holoptic eyes* which touch each other dorsally. The wings are powerful and the abdomen is truncate. It lacks wax glands, pollen collecting apparatus and the sting. The 9th sternum has two claspers and a membranous *aedegus*. The drones are developed parthenogenetically, in a large special cell of the hive. Their number is less in the hive sometimes only one drone in a hive.

Digestive System

The digestive system includes alimentary canal and digestive glands:

The alimentary canal comprises a *mouth* which opens into the *pharynx*. Pharynx is a muscular widened tube that opens into long slender *oesophagus*. The oesophagus running clear through the thorax and open into an enlarged, thin walled sac, the *crop* or *honey sac*. The honey sac lies in the abdomen and it narrows posteriorly and opens into the *stomach*. The honey sac is of special interest in worker. The nectar collected from the flowers is held in it. From the sac the necter is regurgitated into the cells of the comb or given up first to another bee. The upper end of the *stomach* sticks up into the lower end of the honey sac as a small cone with an 'x' shaped value in the summit. This opening is called stomach mouth and is guarded by four lips that take whenever food the stomach requires from the honey sac. Stomach leads into small *intestine* having a circlet of 100-125 slender *Malpighian tubules*. Intestine opens into terminal part of alimentary canal, the *rectum* which opens out through *anus*.

The *salivary glands* are located on the back of the head in the front part of the thorax and open upon the upper part of the labium.

Circulatory System.

It is similar to that present in cockroach. The *heart* is consists of four chambers instead of 13 chambered heart of cockroach.

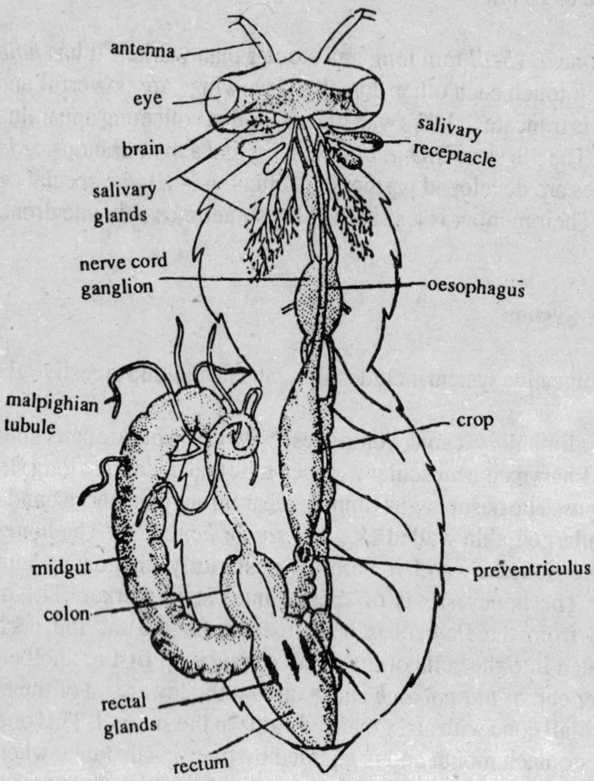

8.6. Digestive and nervous systems.

Respiratory System

The respiration is brought by the tracheal system as found in other insects. In all these are 10 pairs of spiracles present in thorax and abdomen. Two lateral tracheal trunks are enlarged in abdomen to form *air sacs*.

Nervous System.

It is made up of brain, circum oesophageal connectives with double ganglionated nerves cord. There are three thoracic ganglia. Last two are fused. 5 - abdominal ganglia are also fused in a mass.

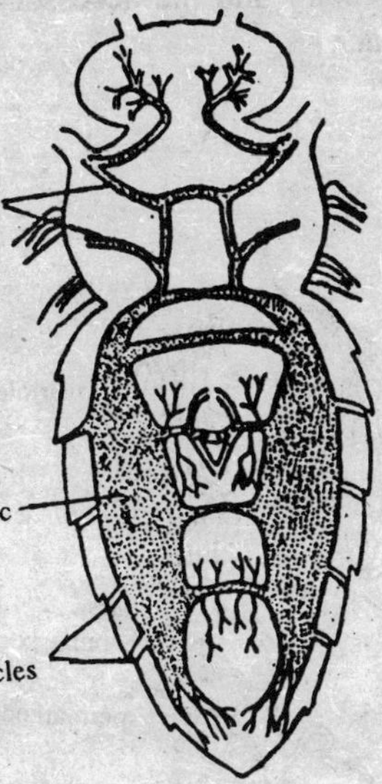

8.7. Apis Respiratory System.

Reproductive System.

The Sexes are separate. The male reproductive organs comprise a pair of *testes*. Each testis is enclosed in a double mebrane and consists of 250-300 seminiferous tubules. From each testis arises a vas deferens. The *vasa deferentia* enlarge to form *seminal vesicles* which are usually

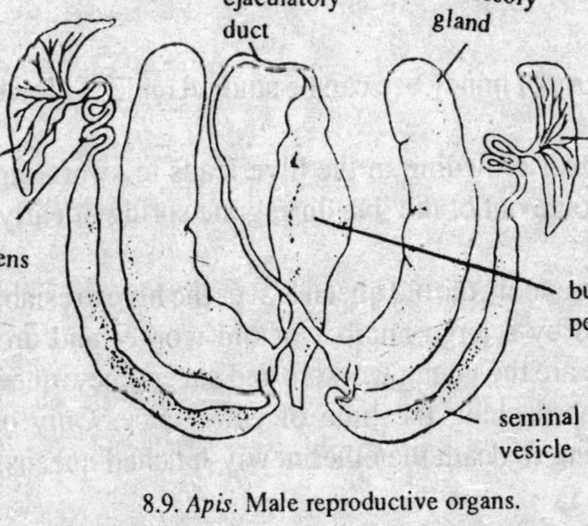

8.9. *Apis*. Male reproductive organs.

cylindrical or sac-like in appearance. The two ejaculatory ducts leaving the vesicles are rudimentary and the accessory glands open into the common *ejaculatory duct.*

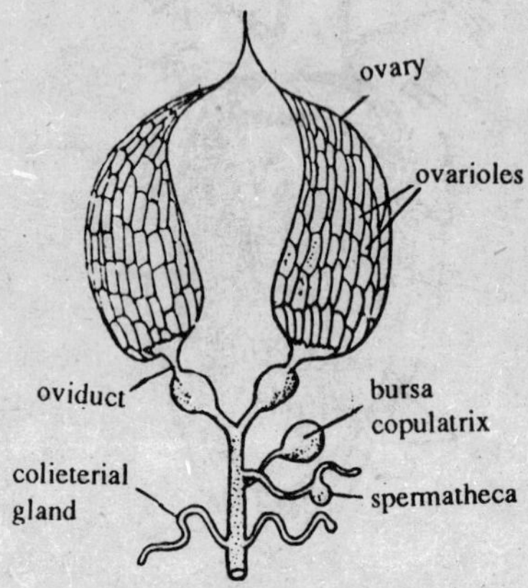

8.9. *Apis*. Female reproductive organs.

The female reproductive organs include a pair of *ovaries*. Each ovary bears numerous ovarioles. From each ovary arises an *oviduct*. The two oviducts unite to form the *vagina* which dilates posteriorly to form *bursa copulatrix.* A median *spermatheca* is present together with a pair of *collaterial glands* which open into the spermatheca.

Life History

The life-history of honey bee can be studied under following heads:

Swarming. Over crowding in the hive leads to swarming. A large number of bees come out of the live during the spring or early summer.

On the fine forenoon, the old queen leaves the hive to establish a new one, accompanied by a large number of old worker and drones. Left behind in the hive are the young workers and several new queens still in their cells but approaching the time of emergence. Only one queen survives by stinging to death the other newly-hatched queens.

Supersedure. In supersedure the old queen with reduced egglaying capacity is replaced by a new, young and vigorous queen. The old and new queen may continue to live together or the old one may be dragged out.

Absconding. Under unfavourable conditions or due to certain enemies the entire colony of bees migrates to a suitable place. The migration of the entire colony is known as *absconding*.

Nuptial Flight

About a week after emergence, the new queen takes her first aerial flight followed by a swarm of drones. The queen flies very high and the drones gradually drop out of the race. The last drone left in the race, mates with her. *Mating* occurs in mid-air, during which the queen receives spermatophores from the drone. The sperms are stored in spermatheca or sperm-reservoir of the queen to fertilize her eggs as long as she lives. The *prime swarm* is led by the old queen while the second *swarm* is accompanied by the newly-emerged virgin queen.

The queen after mating returns to the hive and does not leave it till become old. It was thought that a queen mated only once in its life. But it has now been known that out of experimented 110 queens, about 55 mated twice. This experiment was carried out very successfully in U.S.A.

After the swarming period the remaining drones are pushed out of the hive by the workers, if they survive.

Eggs and Egg-laying- Egg laying starts 3 days after copulation. The eggs are of two types-fertilized eggs developing into queens and workers and *haploid eggs* developing into drones, and are laid into separate chambers or cell. The fertilized eggs are laid into workers or queen cells, while the unfertlized into drone cells. In a season of abundant nectar, the queen lays upto 1000 eggs per day. The eggs are elongated and pinkish and are attached to the bottom of cells.

Larvae. After three days a tiny larva or grub is formed from each egg. There are no eyes and legs in the larva. For the first days all larva are fed on '*royal jelly*' which is secreted by pharyngeal glands of young workers. Afterwards, larvae of the drones and workers are fed on honey

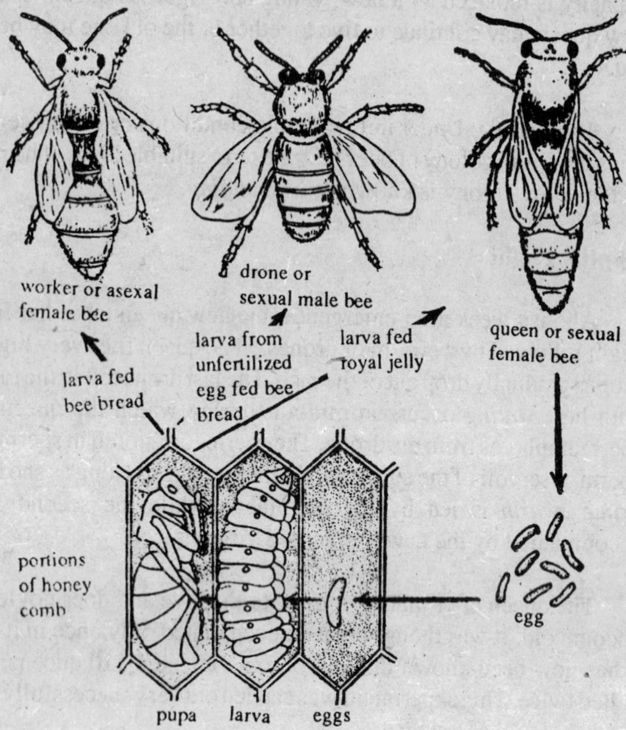

8.10. *Apis*. Life Cycle

and pollen, but the larvae of queen are continuously fed on 'royal jelly'. In this way the food-supply causes them to develop differently. After several moult the larva grows, then its cells sealed with a wax-cap.

Pupa. The larva changes into a pupa in the sealed chamber.

The wax-lid is porous and allows exchange of air for respiration. The pupa secretes a thin silken cocoon around itself and undergoes metamorphosis. It involves a change of legless, wingless and eyeless worm-like creature into a winged insect with legs and eyes. After the pupal period is over, the lid closing the cell is cut off by the young bee with jaws to become free.

The time of development for each caste is different which can be shown as:

Caste		*Egg*	*Larva*	*Pupa*	*Total*
1.	Queen	3days	$5^{1/3}$ days	7days	$15^{1/2}$days
2.	Worker	3"	5"	13"	21"
3.	Drone	3"	6"	15"	24"

The queen produced from her mandibular gland a special secretion called *queen substance.* It is picked up directly by a few workers that are in contact with the queen. These workers spread it to the other. The "queen substance" inhibits the workers from building brood cells for new queens.

The freshly emerged workers are first entrusted with the indoor durites for two to three weeks during which they act as nursing bees, dance attendance on the royalties, look after bood cells, build and repair the comb. Later on, they are put to outdoor duties and they are completely occupied in collection of nectar and pollen, guarding the hive, air conditioning, temperature regulation and ripening the honey, etc.

At the approach of winter, the drones are driven out of hive by the workers. They are either stung by workers or die of cold or hunger.

The queen and workers live there to pass over the unfavourable condition feeding over the stored honey and pollens. On the return of spring the activities are regained.

Communication.

The communication in insects is achieved by many ways such as visual stimuli, sound, chemical stimuli, body secretion etc. The honey been has surprised by their peculiar movement, better known as *'bee dance'*. This peculiar sense of understanding and communication among bees his led *Carl Van Frisch* to be honoured with Noble Prize in 1973.

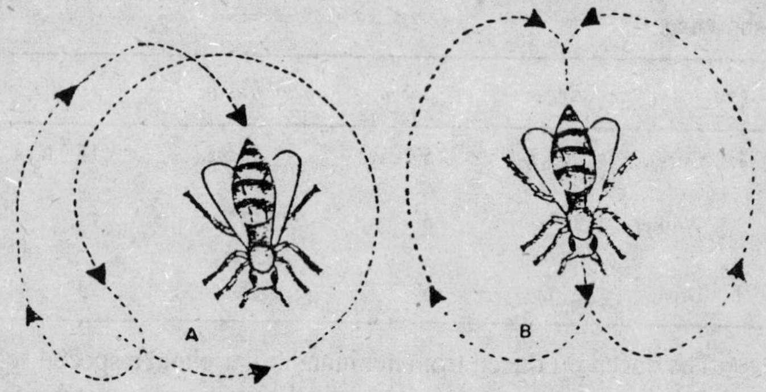

Fig. 8.11. Two different Types of dances. A = round dance B = Tail wagging dance

All the bees of a hive do not go out for the search of food. These are certain workers generally called *foregers* or *scout bees* who perform this duty and on their communication other bees move to the feeding place. These foragers collect the nector and return back to the hive and communicate the source of by performing the dance. The dance is performed in two manners, *round dance* and *tail wagging dance.* if the food source is less than 100 metres from the hive, it performs a çircular dance. It the source is more than 100 metres, the dance performed is making a figure of 8.

The round dance convey the information about the source of food only where as the tail wagging dance indicate the source, distance direction and the quantity of food. Between each semicircle, the bee moves along the vertical plane wagging its abdomen side to side. The frequency of wagging decreases with the increase of dance. The movement of sun has a definit relation with the direction of dance.

When forager performs its dance the other workers join the dance for a while and than leave the hive in the direction of food. *A. Wenner* (1964) claimed that foragers communicated to the other workers of their hive not only by their dance but also by producing peculiar sound which results due to the virbation of their wings. The intensity of sound is proportional to the distance of food from hive and also to the quantity of the food.

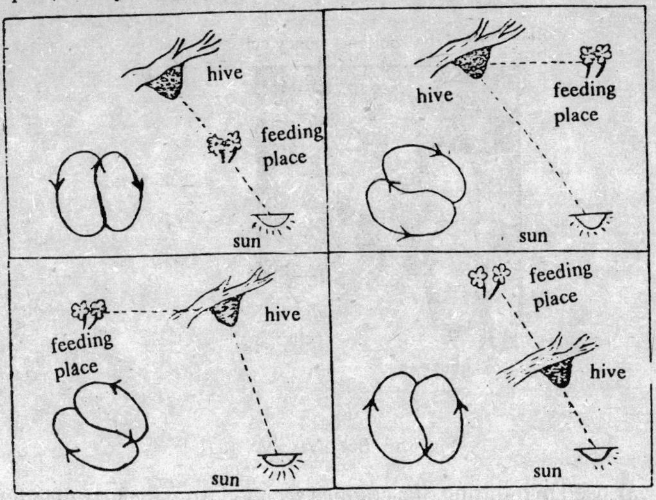

Fig. 8.12. Directional changes in bee dance to devote the feeding place with relation to the position of sun.

Structure of a hive

The highest degree of nest construction among insects is found in bees. The hive is generally built hanging from a building, rock or branch of tree. It consists of a thick vertical sheet bearing two layers of hexagonal chambers known as *cells*. The cells are of following types:

(i) *Storage cells.* These cells contain honey and pollen. These are generally built on the margin and at the top of the comb.

(ii) *Brood cells.* The brood cells contain young stages and are built in the centre and the lower part of the comb. The brood cells are further divided into three types:

(a) *Worker cells.* These are of the size of the storage cells and are used for rearing workers. They lie horizontally in the hive.

(b) *Drone cells.* These are about 6 mm. wide and are used for bringing up the drones. They are also horizontal in position.

(c) *Queen cells.* They are still larger and are used to rear the queens. They may be cylindrical or vase-like in form, and are vertical in position.

There are no special cells for adults, as they keep generally moving about on the surface of the comb or clustering cells which are situated on the upper part of the comb.

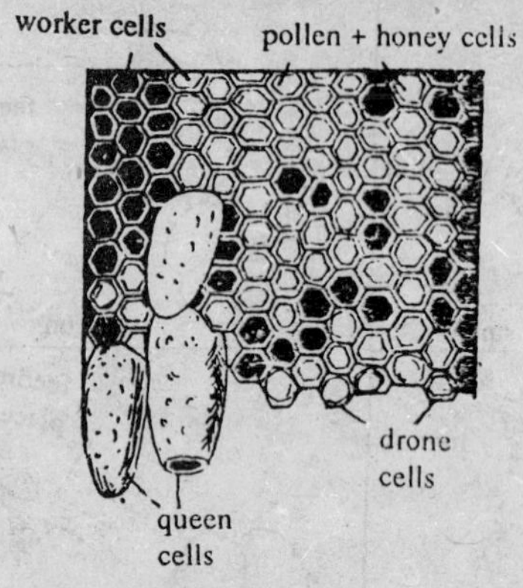

Fig. 8.13. Bee hive.

The wax used in building of a comb is secreted from the wax glands present in the abdomen of workers. This wax has a melting poing of 140°F. Before use the wax is masticated and mixed with secretions of cephalic glands to convert into a plastic substance. The hive is polished with substance, *propolis*, prepared from pollen, is used for making the comb water proof.

Economic Importance

Honey bees are most valuable domesticated animal of man and have been associated with human race from the time of Vedas and Ramayana. They provide us honey and wax and more over they help in pollination.

Honey. The honey is used by man as the source of natural sweets for preparing cakes, breads etc. It has great food value as one pound of honey is equal to 4.5 pounds of grapes or 5 pounds of apples or 3.5 pounds of potato. It is also a very powerful tonic and can be easily compared to 365 UG - thiamin (Vitamin B_1), 268 UG Riboflavin (G), 18 MG Ascorbic acid (C), 254UG Pantothenic acid (B_3) or 0.60 MG Nictonic acid (B_5)

The chemical composition of one pound of honey is: $6^{1/2}$oz *Levulose (fruit-sugar)*, $5^{1/2}$ oz. *Dextrose (Glucose), 9 gm. Sucrose, 3 oz Moisture, 7 gm. Detrines* and gums 1 gm. of Fe, Ca, Na etc. and about 4% of undetermined substance. The colour of the honey and chemical composition and flavour usually depend on the flowers visited by worker-bees.

Bee Wax. The wax is removed from the bee hives and is used in making polishes, candles, corbon papers, electrical and other products.

Pollination. Perhaps the greatest survive of honeybee is the pollination of flowers of fruit plants and seed crops.

Bee venom. It is secreted by poison-glands of stings. It is transparent and, contains formic acid, histamine, tryptophane, sulphur, many proteins, volatile oils, enzyme etc. Clinically it is used in the treatment of rhematism, keratoconjunctivities, iritis, iridocylitis and in the control of blood pressure.

9

CIMEX (BEDBUG)

The bedbugs are domesticated insects and are intimately associated with man throughout the ages. They are sanguivorous as they suck the blood of man. Bed bug includes 30-40 species of wingless brownish blood sucking insects, mostly parasitic on birds and bats. Three species attack man these are *Cimex lectularius, C.hemipterus* and *C. Boueti*. Out of these first and second are more common. The third one is found in Africa, Southern Asia and West Indies.

SYSTEMATIC POSITION

Phylum	-	Arthropoda
Class	-	Insecta
Subclass	-	Pterygota
Division	-	Exopterygota
Order	-	Hemiptera
Genus	-	*Cimex*

Habit and Habitat

They are cosmopolitan in distribution. They inhabit dark, damp human dwellings like houses, hotels, rest houses, barracks, etc. They live in cracks in the walls and floor, in crevices in the beds and furniture, under mattresses, carpets and wall paper, and in similar places.

The bedbug is a temporary ectoparasite of man feeding on the blood. They are nocturnal, but may come out during the day. A well fed bug can do without food for several months or even a year. The starving bug may resorts to cannibalism. They are oviparous and undergo gradual metamorphosis. This bad smelling pest lives for 5-6 months.

Morphology

Shape, Size and colour. The body is oval, dorsoventrally flatened, measuring about 5 mm in length and 3 mm in width. The colour of the body is brown becoming deep purple or red after it has fed.

Division of body. Like other insects, the body is divisible into three parts:

1. Head
2. Thorax
3. Abdomen

1. *Head.* The head is short, broad and fits into a notch of the prothorax. It bears a pair of compound eyes, 4 jointed antennae, and mouth parts. The simple eyes or ocelli are absent.

The *mouthparts* are of the *piercing* and *sucking type* and form a hollow *beak* or *proboscis* in front of the head. The beak is formed by the labium, which bears a middorsal groove for enclosing the four sylets of two mandibles and two maxillae. The saw like mandibular styles are used for making puncture in the skin whereas the maxillae for the inward flow of blood and downward flow of saliva. This is facilitated by two channels of the maxillae. The maxillary and labial palps are absent.

At rest, the restrum is bend beneath the throax but while feeding, it is pointed downwards or forward. The mandibular sylets make a puncture in the skin and the saliva runs down the puncture and prevents the clotting of blood.

2. *Thorax.* The thorax is made up of three segments prothorax, mesothorax and metathorax. Prothorax is very large; Its pronotum is notched anteriorly to receive the head. Mesothorax is very small.

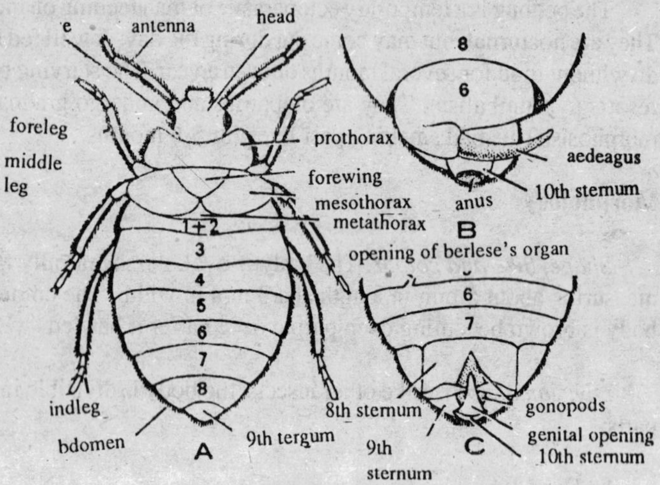

Fig. 9.1. *Climex.* A = Adult man, B = Genitalia of male of male, C = Posterior end of Female

Metathorax is covered by a pair of vestigial *fore-wings* or *hemelytra* that arise from the mesothorax. There are no hind-wings. Each thoracic segments bears ventrally a pair of short, stout legs. The legs have 3-jointed tarsi and strong claws. Stink glands open on the ventral side of the metathorax.

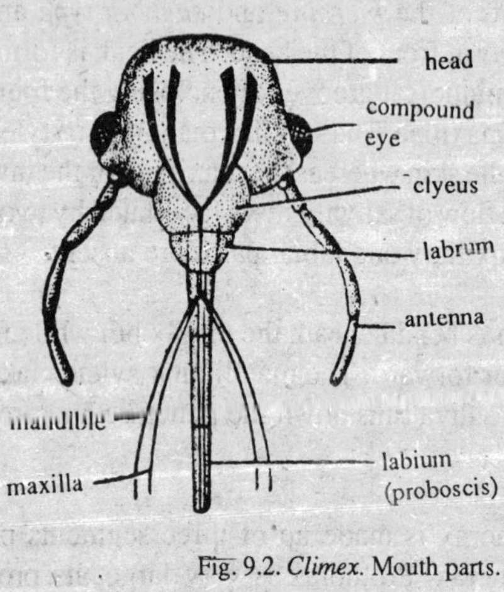

Fig. 9.2. *Climex.* Mouth parts.

3. *Abdomen.* The abdomen is flat and made up of 7 segments. In the male, it is relatively narrow and pointed and ends in a curved hook-like

clasper, which serves as a sheath for the *aedagus* or *penis*. In the female the abdomen is broad and rounded and has a prominent notch in the 4th sternum a little to the right of the middle line. The notch is the opening of a peculiar copulatory pouch, the *organ of Berlese* or *spermatophagous organ* the cells of which are thought to ingest and destroy the excess sperms.

Life History

Oviposition. After copulation the female lays eggs in holes, cracks, cervices etc. A female depositing 200 to 500 eggs in batch; the egg is 1 mm long, yellowish-white in colour, it is barrel-shaped with a raised lid at one end containing a micropyle; eggs are laid throughout the year in warm countries.

Nymph. In about 6 to 10 days duration the young bug or nymph hatches out of the egg by pushing off the lid. The nymph is about the size of a pin head and is semitransparent. In its general appearance, the nymph resembles the adult except that it is smaller and paler.

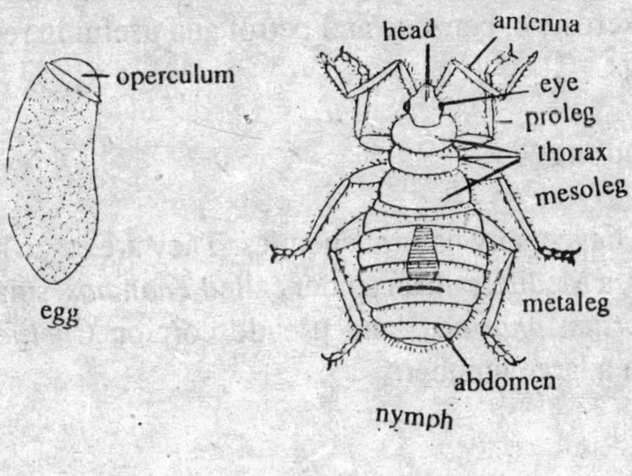

Fig.9.3. *Climex*. Mouth parts.

Few hours after hatching, the nymph becomes capable of piercing the host skin and sucking blood. It searches for man and feeds to its capacity in 3-4 minutes. It consumes blood three to four times of its own weight. If man is not available, it feeds on the blood of older bugs. After feeding it becomes rounded and bright red in colour. Now it settles down and in one to four weeks it moults into the 2nd stage nymph. The duration of 1st stage nymph depends upon temperature and other factors. The nymph undergoes five moults before changing into the adult. On the whole they take about 29 days to complete their like history. Therefore, the development of bed-bugs is direct, *i.e. without metamorphosis*.

Economic Importance

The bedbug is a serious household pest. They feed on mammalian blood causing inflammation and irritation. This is due to the introduction of saliva into the human blood during feeding on it.

It is believed that that the bedbug act as a transmitting agent of *kala azar* (caused by *Leishmanisa donovani*), *relapsing fever* (spirochaetosis), typhoid etc. They, however, do not cause the outbreak of these diseases in an epidemic form.

Control

The best method, to control them, is the spray of DDT (5% in kerosene). This can be sprayed on all the walls and furniture where the bugs rest. Fumigation with sulphur, sprinkling boiling water, spraying of emulsion of kerosene, benzene and petrol and useful in reducing infestation bedbugs.

Enemis of bedbugs

Bedbugs have many natural enemies. They are eaten by large bugs like *Reduvius*, a Mediterranean spider called *Thanatos*, small red house-ant (*Monomorium pharanis*) and pseudoscorpion *Chelifer*. Man also destroys them a large number:

10

TERMITES

Termites commonly called whiteants are social and polymorphic insects living in colonies. They have a well set caste system.

SYSTEMATIC POSITION

Phylum	-	Arthropoda
Class	-	Insecta
Subclass	-	Pterygota
Division	-	Exopterygota
Order	-	Isoptera.
Type	-	Termites

Habits and Habitat

The termites occur throughout the tropics and temperate countries. About 1600 species are known. The most common are *Microtermes* and *Odontoermes*. These are colonial insects. They either burrow in wood or make substerrancan nests or *termitaria*. They feed mainly on wood and the cellulose is digested with the help of symbiont flagellate, *Trichonympha campanula* living in the intestine of termites.

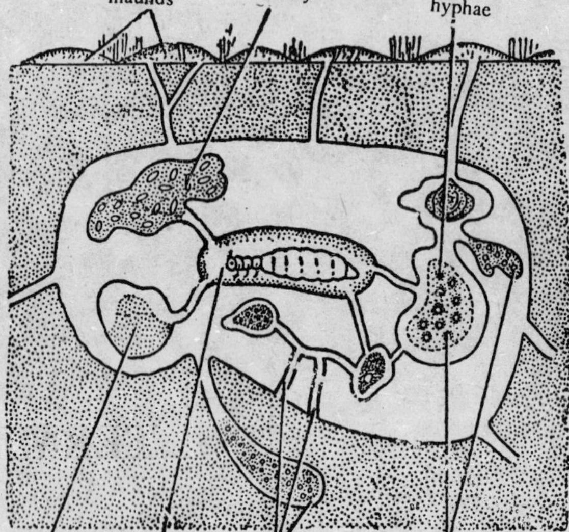

Fig. 10.1 Termite nests.

External Features

These are soft-bodied small medium sized insects. The colour of the body is white or pale yellow. The body is elongated and flattened with smooth or hairy surface. The body is divisible into three parts.

1. Head
2. Thorax
3. Abdomen.

1. *Head*. The head is free and well selerotized. It bears a pair of short or long beaded antennae. The compound eyes may be well developed, vestigial or absent similarly ocelli may or may not be present. The mouth-parts are chewing type.

2. *Thorax*. It is divisible into three segments: prothorax, meso-throax and metathorax. Each segment bears a pair of legs on ventral side.

Each leg with 4 or 5 jointed tarsus and a pair of claw. The fertile form bears two pairs of wings on the dorsolateral side of mesothorax and metathorax. The wings are membranous and extend behind the body. They can be shed at basal suture. The wings are absent in sterile forms.

3. *Abdomen.* The abdomen bears 10 segments, ending in a pair of short 2-6 jointed *cerci.* In male the 9th segments bears a pair of *anal styles.*

Polymorphism

The termites are polymorphic and includes two forms:

1. Fertile caste or reproductive caste
2. Sterile caste

1. *Fertile Caste.* These are also called reproductive or sexual castes. The fertile individuals comprises the following forms:

(i) *Large winged or Macropterous forms.* These are true king and queen. Their colour is yellow, brown or black. The wings extend for beyond the hind-end of the body. The compound eyes are widely separated and are large. The ocelli are usually present. They live in royal chambers. The sole duty of king is to fertilize queen and queen is to lay eggs.

(ii) *Short winged or Brachyterous forms.* These are supplimentary reproductive individuals. They have a less chitinized body of pale-yellow colour. The two pairs of wings are vestigial, short and stump --like. The compound eyes are small and the ocelli may or may not be present. These forms replace the true king and queen in case the king and queen die.

(iii) *Wingless or Apertous forms.* These are called *ergatoid* king and queen. Their structure is similar to that of a worker. Their body is colourless, wings are absent. The compound eyes are vestigial and the ocelli are absent. The sex organs are poorly developed.

2. *Sterile Caste.* These are wingless forms with rudimentary reproductive organs. These are blind or eyeless and colourless. The sterile caste is of three types:

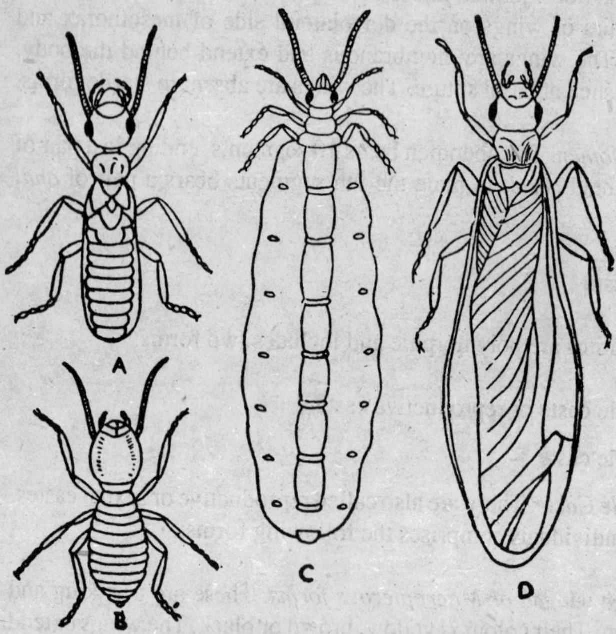

Fig. 10.2. A= Queen of termites without wings after nuptial; B= wingless sterile worker; C= mature queen; D=the winged male before nuptial.

(i) *Workers*. These are the most numerous of all castes. They are generally *dimorphic,* i.e., occur in two forms: small and large. In some cases they are *trimorphic,* being small, intermediate and large. The workers construct and repair the nest: collect and distribute the food: look after the eggs; and feed the nymphs, soldiers and the royals couples (true kings and queens). In some species, the workers cultivate a certain fungus on a finely chewed vegetable matter for feeding the royal couples and nymphs.

(ii) *Soldiers*. The soldiers are the most highly specialized. They defend the colony against enemies. These are pale coloured, pigmented forms with large chitinized big projecting mandibles. The brain is small and compound eyes are vestigial. In certain species there may be three grades of soldiers: large, medium and small sized.

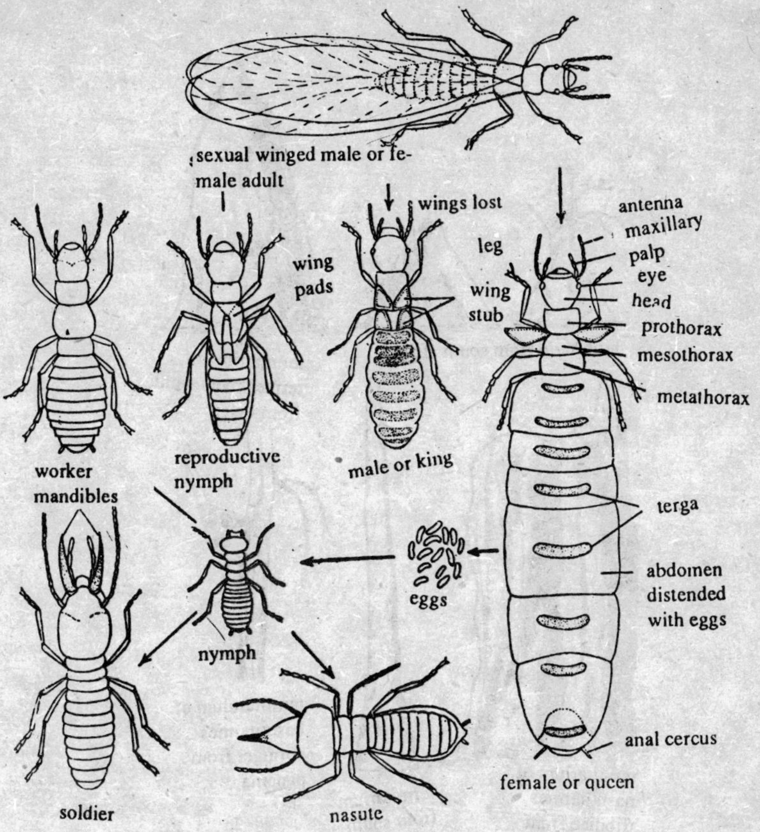

Fig. 10.3. Different castes and life history of Termite.

(iii) *Nasutes*. In some higher genera the mandibulate soldiers are replaced by *nasutes* or *proboscideans*. These forms are short with vestigial mandibles. Their head is prolonged into a rostrum, which bears at its tip the opening of *frontal gland*. They throw a jet of sticky secretion of the frontal gland on the invaders through their rostrum. This secretion immobalizes or kill the enemy. The secretion is also used to dissolve hard substances like stones, concrete which are in the way of the workers when building. They may be dimorphic or trimorphic i.e. small, medium and large sized.

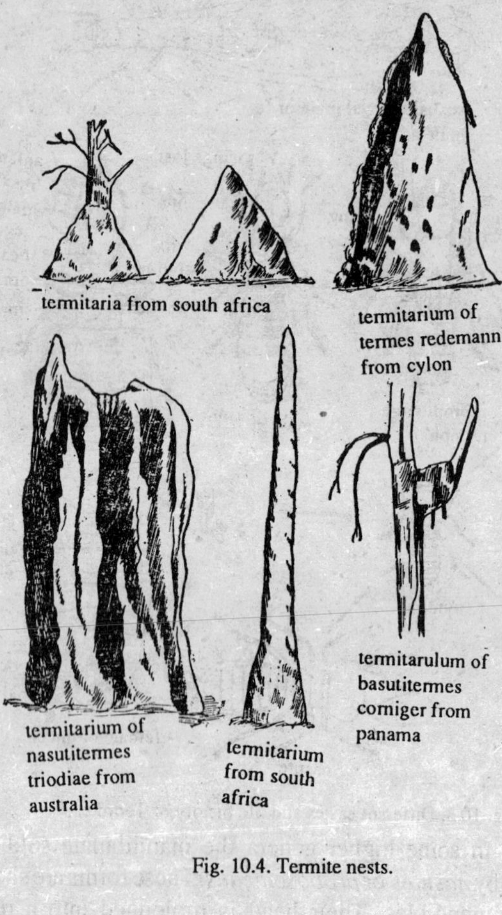

Fig. 10.4. Termite nests.

Nests

Primitive termites live in galleries bored in wood, logs, fenes, poles, furnitures etc. Some termites make tunnels in the ground, they destroy roots of grasses, vegetation and crops. Other construct a *nest* or *termitarium*. The nests consists of a labyrinth of irregular tunnels and chambers and may be entirely sub-terranean or with a mound above the surface. These mound may be from few metres to nine metres. The natives of Belgium and Congo clear them of the termites and use them as their dwelling huts. There are other types of the termites which form termitaria of cartn and masticated wood. These are of the size of a football and are placed upon the trees.

Life History

Swarming and mating. In the rainy season, with the first shower of rain, a large number of macropterous forms (kings and queens) leave the termitarium through the holes made by soldiers and fly away to new sites. It occurs during night or day depending upon the species. Mortality is very high during and after swarming. A large number of these forms are eaten up by lizards, birds, toads, ants etc. Only a few lucky ones survive and found a new colony. At the new site the king and queen mate. Thus the flight is not a true *nuptial flight* but rather a *dispersal flight*.

The king and queen together excavate a small burrow in the ground called the *nuptial chamber*. The queen lays eggs. These eggs are tended by the royal couple and hatch into workers, which soon take up their duties. The queen now undergoes drastic modifications. Her ovaries and fat bodies increase greatly while the jaw muscles degenerate. The king lives with the queen for life, and they mate at regular interval. The queen grows her abdomen to an enormous size, 50-75 mm. Her legs being absurdly small, she is unable of the slightest movements and she lies passively in the royal chamber. She is tended by the workers and defended by the soldiers. On an average she lays 4000 eggs per 24 hours. The queen may live for 6 - 15 years and lays about a million eggs during her life.

Development. The eggs are pale, smooth, oval or elongated. They hatch into wingless nymphs, adults excepts in size., wings are absent and sexually immature. After several moults the nymph reaches the adults stages. Formation of different castes from similar eggs in perhaps determined by variable food supply, hormones, pheromones, social beahaviours etc.

Economic Importance

The termites are highly destructive insects. They attack and damage all sorts of plant materials, like wood used in buildings and bridges, furnitures, railway-sleepers, fences, crops, books and papers.

However, they are useful also in certain respects. They increase the fertility of the soil by making it porous with their burrows like the earthworms. They dispose off dead wood in the forests to make room for new vegetation. They are eaten by natives in some tropical countries, like South East Africa.

Control

They can be destroyed in the field by pouring kerosine and cerbon bisulphide and chloroform or petrol. Nests can be fumigated by Arsenic, Sulphur and Kerosin oil. Trees must bee banded by coal-tar or 5% DDT emulsion in oil should be painted on the trunk to 0.45 metre from ground level. This treatment should be done after mansoons. A colony of termites can be eradicated only after findings out and killing the queen.

11

BOMBYX MORI

Bombyx mori commonly called as *Chinese* or *mulberry silkworm moth* is a native of China. It is well known for the pure silk. The moths are reared for silk.

SYSTEMATIC POSITION

Phylum	-	Arthropoda
Class	-	Insecta
Subclass	-	Pterygota
Division	-	Endopterygota
Order	-	Lepidoptera
Genus	-	*Bombyx*
Species	-	*mori*

Types of Silkworms

1. *Mulberry silkworm.* The mulberry worm (*Bombx mori*) belongs to the family Bombysidae. It produces most of the raw silk we use. This silk is supposed to be superior in quality to the other types due to its shining and creamy white colour. The catterpiller feeds on mulberry leaves.
2. *Tasar silkmoth.* A number of species of genus *Antherae* produce tasar silk. These are *A. mylitta, A. paphia, A. royeli, A. peryni* etc. Their catterpiller feed on leaves of Arjan, Asan, Sal, Oak and various other secondary food plants.

3. *Eri silkmoth.* The eri silk is produced by *Attacus ricini*. Their catterpiller feed an castor leaves. The silk is creamy white but is less shining.
4. *Munga silkworm.* The munga silk is obtained from the catterpiller of *Antheraea assama* which feeds an Champa, Cinnamon etc.

The mulberry, silkmoth never occures in the wild state, it is completely domesticated. *Bombeyx* is extensively cultivated all over the world and usually produces a single brood in a year i.e. *univoltine* . Some strain, however, produce 2-7 broods a year (polyvoltine) and are cultivated in warm climate.

External Characters

The adult moths are robust and cream white in colour with several lines across the fore wings. The body is very heavy and very hairy. It is about 25 mm long and has a wing expanse of 40-45 mm. Though both the pairs of wings are apparently well developed the moths fly very rarely. Females are longer than the males. The proboscis is absent, antennae are bipectinate or feathery in both the sexes. The life-span of adult is short usually ranging from 2-3 days. It does not take food during this period, only copulates.

Life history

Male and female moths copulate in tail to tail position. Life-history comprises four stages.

1. Eggs
2. Larva
3. Pupa
4. Adult

The metamorphosis is complete.

1. *Eggs.* Soon after fertilization, each female lays about 300-500 eggs in clusters upon the leaves of mulberry tree. They are fixed to leave with a gelatinous secretion. The eggs are small, smooth and spherical when laid they are yellowish white but become darker later on. The female dies within 3 or 4 days after egg laying. In the univoltine race

overwritering takes place in the egg stage, as such the eggs may hatch in months. In polyvoltime race the eggs hatch in a few days.

2.*Larva*. The larva is called *caterpillar or silkworm*. The catterpillar larva which hatches from the egg measures 5-7 mm in length. These caterpillars move on the leaves in a characteristic looping manner. Their body is rough, wrinkled and greyish in colour. They are made up of 12 segments which is distinct into three parts: head, thorax and abdomen. The head bears mandibulate mouth parts with which they feed upon the leaves. The thorax is 3 segmented and all the segments bear a pair of true 5 jointed legs ventrally. The abdomen is 10 segmented and bears 5 pairs of unjointed, stermpy *prolegs* or *pseudolegs*. These prolegs are present in each segment 3rd, 4th, 5th, 6th and 10th. A short dorsal *anal horn* is present on 8th segment. The prolegs of on the last segment turn backwards to form the *claspers*. There is a row of small respiratory pores, the *spiracle*,on either side of the abdomen.

The larva has only three pairs of simple eyes or *ocelli*. The antennae, maxillae and labial palps are rudim, entary.

The larvae feed voraciously upon the mulberry leaves and grow very quickly. They stop feeding, become inactive after 4-5 days, and than 1st moulting takes place. The 2nd stage larvae resemble the 1st stage larvae except that they are slightly bigger is size. They also eat voraciously for next seven days than 2nd moulting takes place and third stage larvae are formed. The larvae repeats this process for four times. The maturity is achived in about 45 days since the time of hatching. The matured caterpillar now measures 17-10 cm in length. By this time the formation of a pair of salivary glands is completed. Since these glands secrete silk they are called *silk glands*.

3. *Pupa*. The mature caterpiller stops feeding and excrete its last excreta and begin to secrete the stickysecretion from the silk glands through a very fine pore celled *Spinneret* situated on the hypopharynx. The secretion is continuous and after coming in contact with the air the sticky secretion is converted into a fine, long and solid thread of silk. The thread is made up of five filaments stuck together by *sericin* secreted by two other glands. The thread becomes wrapped around the body of larva forming a *cocoon*. This process continues for 3-4 days. The cocoon

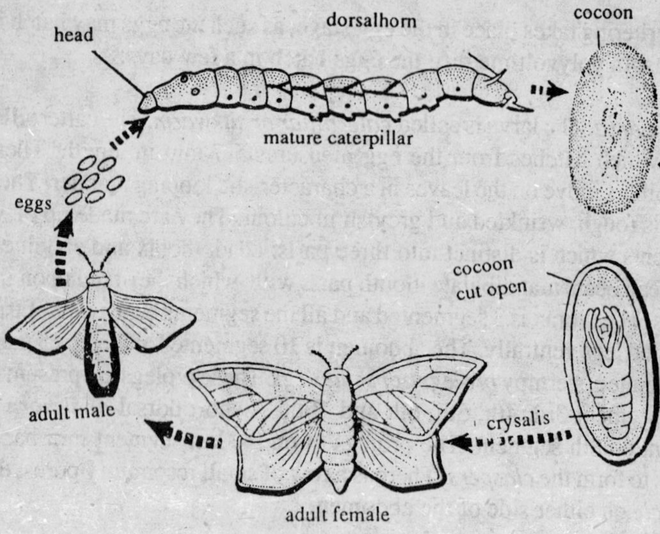

Fig. 11.1. *Bombyx mori*. Life history

serves as a comfortable house for the protection of the caterpillar for further development.

The cocoon is a thick, white or yellow and oval capsule. The outer or initial filaments of the cocoon are irregular, but the inner one that form the actual bed of the pupa is one long continuopus thread about 300-450 metres long wound in ring. The silk thread is secreted at the rate of 150 mm per minute. Within 15 days the caterpillar is transformed into a brownish *pupa* or *chrysalis* . The pupa lies dormant but undergoes metamorphosis. The larrval organs such as prolegs, claspers, anal horn and mouth parts are lost. The adult organs develop. Within 12-15 days the adult is formed.

4. *Adult (Imago)*. The adult secretes an alkaline fluid that moistens and softens on end of cocoon. The cocoon breaks at this end and the adult comes out. Soon after the emergance, the silk moths mate, lay eggs and die.

Economic Importance

The silkmoth is a very valuable insect as it gives us silk. Besides silk it gives *gut*.

Sericulture.

Sericulture is the production of silk by rearing silkworms. The wilk industry originated in China about 3000 B.C. A Chinese Princesses Siling Shi was the first to discover the art of receling an unbroken filament from a cocoon. This art was kept a close secret for nearly 3000 years. This art later on spread to the rest of the world.

Sericulture is a regular industry. India has silk-producing centres in Assam, Bengal, Madras, Mysore, Punjab and Kashmir. Healthy eggs of high-yielding strains are procured from sericulture research stations. The eggs are placed in paper-lined trays made of spilt bamboo. The trays are kept out of the reach of ants. The eggs are periodically stirred by a feather. Larvae are given chopped leaves of a good quality mulberry trees 5 to 9 times a day. The pupae are not allowed to become adults, as the latter rupture the cocoon to escape, and thereby cut the silk thread of cocoon into pieces. About 10 days after cocoon formation, the pupae enclosed in them are killed by soaking the cocoons in hot water, by steaming them, by giving them dry heat in an oven or in the sun, or by fumigation. Some cocoons are kept safe. Their pupae are allowed to develop into adults to get seeds (eggs) for the next crop (brood).

Reeling of raw silk from cocoon

Reeling constitute an important aspect of sericulture because cocoon production is directly related to reeling industries. Before reeling the thread the cocoons, with dead pupae, are dipped in a container of hot water for more than 10 minutes. During this period they are continuously stirred with a rod. For tasar silk special cooking techniques is employed. The cocoons are dipped in 0.5% Na_2 Co_3 solution for 18 hours. Than cocoons are subjected to steam cooking for $2^{1/2}$ hours. After 24 hour the cocoons are treated with 0.5% formaldehyde solution for 15 minutes. Due to this process, their outer portion is loosened and removed in the form of long tapes and the end of the continuous filament is found. The filaments of several cocoon are picked up and passed through the 'glass eye' on to the reel. The thread thus reeled forms the *raw silk* of commerce. About one kilogramm of raw silk is obtained from nearly 55,000 cocoons and one cocoon yields about 300 metres of silk-thread. The waste superficial threads and damaged cocoons are combed and teased. The resulting fibres are spun and this is spun silk.

Mulberry - silk is restricted to Purnea district, along the border of West Bengal. The total production of silk is of the value of aboput 3.5 crores in Bihar. Munga silk is only produced in Assam.

12

MOUTH PARTS IN INSECTS

Every organism should possess, in the view of feeding, organs which are either primaries or secondaries modified structures those enable the animals to feed and live as long as they are alive. The insects shows various specializations in their mouth parts. No doubt the insects are most successful forms among invertebrates and are adpated to the nature or environment so intimately as that their number is increasing constantly. The success of this group in all respect reveals vivid variations in mouth parts that are also being modified according with their modes of lives.

According to the food and manner of feeding the mouth parts are of following types.

1. Biting and chewing type
2. Chewing and lapping type
3. Rasping and sucking type
4. Piercing and sucking type
5. Sponging type
6. Siphoning type

Out of these the rasping and sucking type is considered to be less important because the parts radiating to the piercing and sucking type.

1. Biting and chewing type

The biting and chewing type mouth parts are considered to be most primitive and they are said to be *mandibulate* types of mouth parts. These mouth parts are found in groups such as Orthopters, Isopters some Coleperans etc. The mecahnism of feeding is simple because that are meant for pinch off, chewing up and swallowing the pieces of plant or animal tissue.

The mouth parts consists of :

(i) Paired mandibles

(ii) Paired maxillae

(iii) Labium or lower lip.

(iv) Labrum or upper lip

(v) Hypopharynx

(vi) Epipharynx

These mouth parts are further divided into three subdivisions:

(a) Grinding type

(b) Grasping or Predaceous type

(c) Mandible- suctorial type.

(a) *Grinding type*. Especially this type is found among cockroaches and grasshoppers. Becasue of being considered them as generalized one so we have to describe the various parts of mouth parts first:

(i) *Labrum or upper lip*. It forms the a roof of the mouth cavity, hanging from clypeus and variable in shape and size.

(ii) *Maxillae* These are paired and attached laterally. Each maxilla consists of a basal part, the *protopodite* which is made up of two segments, *cardo* and *stipes*. The stipe bears two branches distales, the inner *endopodite* and an outer *exopodite*. The endopodite cosnsists of two lobes, inner *lacina* and outer *galea* which bears the taste buds. The exopodite is a branch of stipes possessing 5-jointed *maxillary palp* which acts as tectile organ.

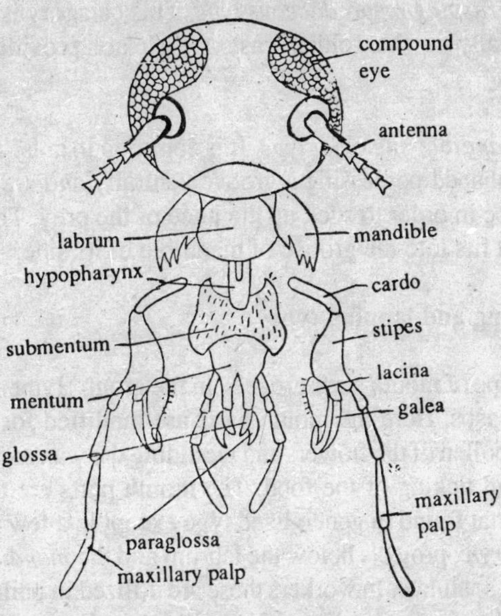

Fig. 12.1. Mouth parts of Cockroach.

(iii) *Mandibles*. A pair of mandibles lie in the mouth cavity and they work transversely to masticate or grind the food. Each one is an unjointed and triangular piece usually, which are very stout. The inner edge of mandible bears tooth-like *denticles* which interlock while capturing the prey. The smooth molar area provides the masticatory surface. Proximal to molar area is a membranous lobe, the *prostheca* which bears sensory hairs.

(iv) *Labium or lower lip.* It forms the floor of mouth cavity. It is formed by the fusion of two maxilla-like appendages. The basal large part is *submentum,* which supports the some-what smaller *mentum*. The mentum supports the *prementum* and paired *glossae* and *paraglossae* together constitute the *ligula*. The prementum bears, on each side, a three-jointed *labial palp* borne on a basal projection, the *palpiger*.

(v) *Hypopharynx.* It is a median process called the tongue, occupying in the buccal caity. It bears at the base an opening of saliwary duct.

(vi) *Epipharynx.* It is a membrane like structure lining the under surface of the labrum and bears taste buds.

(b) *Grasping or predaceous type.* This catagory is found in Isoptera especially in the soldier casts which are provided with strong mandibles.

(c) *Mandible-suctorial type.* It is common in ants. Their mandible are sickle-shaped possessing a groove ventrally and are provided with sucking tube in order to suck up the juice of the prey. The maxillae are slender and fits into the groove of mandible of its side.

2. Chewing and lapping type

This type of mouth parts are seen in the group Hymenoptera such as bees and wasps. Here the mouth parts are modified for collecting the nectar and pollen of the flowers and moulding the wax and serve for both chewing and licking of the food. The mouth parts are the same in all respect as that found in generalized type except in a few modifications. The *epipharynx* projects below the labrum and the *mandibles* are soft or smooth and spatulate. In workers these are utilized in building the honey combs. *Galea* is elongated and blade-like but the maxillary palps are vestigial. The *labium* is articulated with basal triangular *postmentum* or

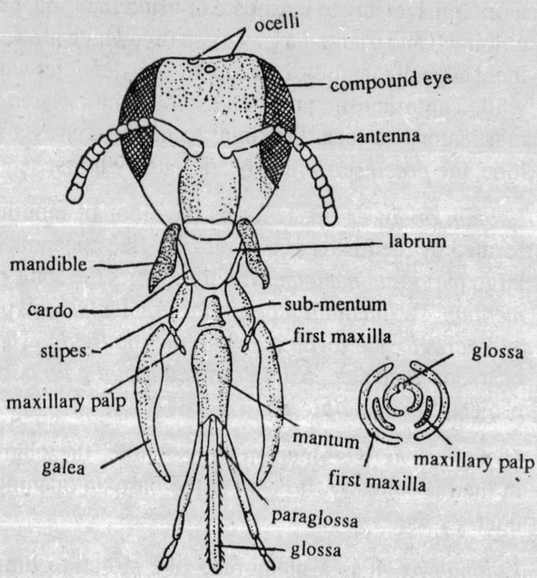

Fig. 12.2. Mouth parts of *Apis*

submentum and *mantum* which is having muscles. *Paraglossae* are greatly reduced where as the *glossae* are much elongated constituting the retractile ligular tongue which terminated distally as a honeyspoon or*labellium*. The labial palps are well developed when the animal wished to take food, the terminal lobes of maxillary galeae and labial palps are brought close together to form a groove or hollow tube which can be inserted into the flower where the labellum moves forwards and back-wards to collect pollen and to suck up the nectar of flower. There are two views either the liquid food can be drawn by automatic capillary action or it is the action of pharyngeal muscles.

3. Rasping and sucking type

Such type of mouth parts are seen in Thrips. The mouth parts are adapted for lacerating and sucking. One of the mandible is reduced markedly but there is nothing like piercing. The stylet is used for escaping the surface of the prey.

4. Piercing and sucking type

This type of mouthparts are common in bugs, mosquitoes, aphids, scale insects, fela, flea and some other forms in which the mouth parts are modified with regards to diet. These are modified in four subtypes:

(a) *Hemipterous type*. Here the four needle -likc stylets, which are grooved lengthwise and lying in the long jointed lower-lip rostrum, are present. The labium is modified into 3 jointed ros-trum. A pair of sharp, pointed mandibles and a pair of slightly shorter maxillae with serrated edges are presemt. These four are modified into stylets and are rectactile and protractile, unlike that those of a mosquito. The labial and maxillary palps are altogether absent. The inner maxillary surface are grooved so as to form two exceedingly fine channels out of which one is feeding canal and the other is salivary canal. The labrum does not take part into the formation of food channel.

The mechanism of feeding is that the beak is bent and piercing organs are stylets which are gliding up and down finally sunk deeply into the flesh of the victim. By the strong motion of the pharynx the blood or juice is sucked up through the tube formed

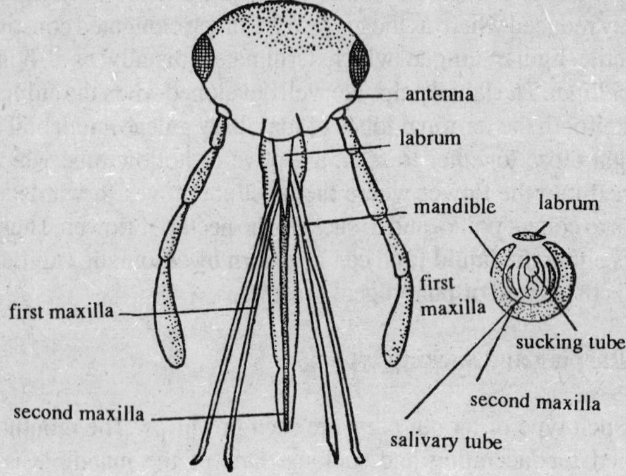

Fig. 12.3 Mouth partsof *Climex*.

of piercing organs. It is found in bedbugs and in some aquatic bugs *(Sphaerodoma)*

(b) *Anoplurous type*. Such type of mouth parts are found in hair-louse in which the mouthparts are withdrawn into a pouch under the pharynx, if not in use. The pouch is said to be style sac, the rest is all the same as we found in bugs. Besides a minute tooth-like structure on the head called *buccal denticle* is present. The latter enable the louse to anchor its head to the skin of the host thereby it starts feeding.

(c) *Flea type*. This type of mouth parts are well marked in fleas. Which are parasites on the domesticated animals like dogs, rats etc. This type is characterized by the presence of a pair of long blade-like mandibles with serrated edges. Labium consisting of two halves fused at their bases so as to form a hollow proboscis enclosing the mandibles, the labrum and the epipharynx, which are fused to cover up the mid-dorsal opening of the labium. These mouth parts are adapted to pierce and suck to the blood of the host.

(d) *Dipterous type*. This subtype is found in mosquitoes. These are adopted for piercing and sucking. The mandibles and maxillae are modified into fine rod-like stylets which are used to pierce the skin of host. The labrum is long needle like and fused with the epitharynx and covers the mid-dorsal opening of the labium. The hypc-pharynx is double edged, served with a mid-dorsal groove, which forms the salivary passage for the secretion of salivary glands. The labial palps are absent, but maxillary palps are present.

While feeding, the hypopharynx and labrum epipharynx become closely applied to enclose a tubular passage for the of ascend of the blood through this tube. The salivary secretion is poured through the groove of hypopharynx in order to check the clotting of blood.

In male mosquitoes, the mandibles are totally absent so that they suck plant juice, whereas the female sucks the blood of human beings and other cattles because of piercing the skin by powerful mandibles.

5. Sponging type

This type of mouth parts are found in Diptera such as houseflies, *calliphora* and others. The adaptability of the mouth parts suited alone to suck as the liquid and not piercing the tissue of enemy. These forms are capable of sucking even soild food which has to be converted into liquid by their saliva before sucking.

The labium is modified into so called *proboscis* which has got a groove with membranous sides and the groove is covered by *labrum, epipharynx* and *hypopharynx* dorsally. The proboscis expands dorsally into a oral-disc or *labellum* which possesses a number of pseudotracheae. Its proximal part is cone-like known as rostrum bearing *maxillary palp* and the middle is *haustellum* with a middorsal groove serving as a food passage. All the pseudotracheae are converging into the central aperture, the *mouth*, which leads into the food passage. The mandibles are absent while the maxillae are represented by two maxillary palps, each made up of a single piece.

Generally the two labellae are kept retracted upwards and backwards during rest, but these labellae are protracted and expanded and placed over the liquid food. The fly can consumes a solid food particle of an easily soluble kind like sugar, the fly, extrude a drop of saliva to make the solid in dissolved form. The liquid food is drawn by the pseu-

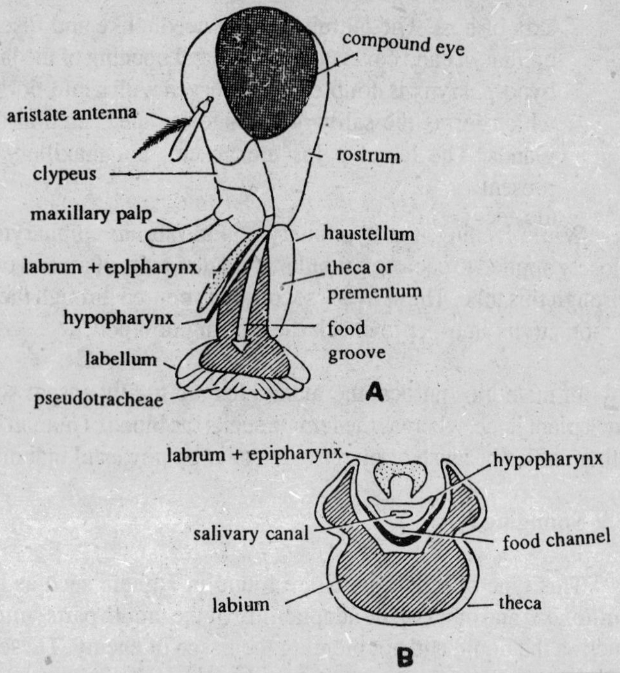

Fig. 12.4. Mouthparts of musca

dotracheae and than passed on to the food-canal which is formed by labrum-epitharynx and hypopharynx lying in the mid-dorsal groove of the haustellum

6.Siphoning type

This type of mouth parts are seen in lepidopterans (butterflies and moths). These are highly specialized for sucking up the juice of flowers and fruits.

The mandibles are usually absent but if present they are very much reduced. The labium is too much reduced. The maxillary palps are considerably vestigial. The labium is represented by a triangular plate bearing the labial palps. The galeae of maxillae are modified into a proboscis through which liquid is drawn up. The proboscis is kept coiled when it is in rest.

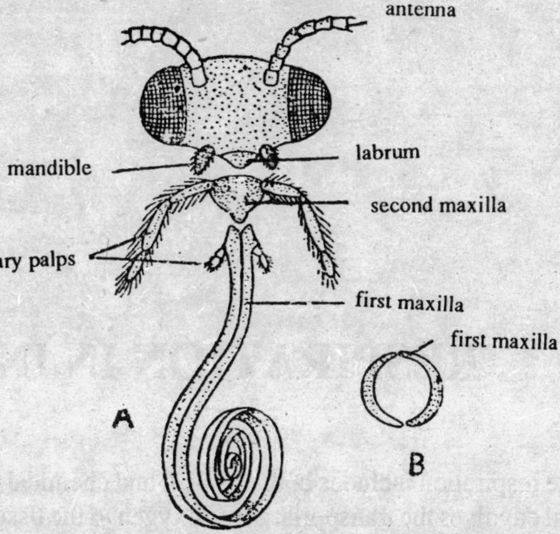

Fig. 12.4. Mouth parts of Butterfly.

When the insect wants to feed than only the watch spring-like proboscis becomes uncoiled and extended to reach the spot where its diet is available.

In Hexapoda the function of the mouth elements is seen in two quite different lines upon which the insects are grouped as the mandibulate possessing the biting and chewing type and other in Haustellata consisting of sucking. From the evolutionary point of view it is decided that the biting and chewing parts are generalized primitive kind from which other types have given rise.

13

RESPIRATION IN INSECTS

The respiration includes both physical and chemical processes. The physical envolves the transportation of oxygen to the tissue and removal of carbon dioxide from tissue where as chemical process envolves the oxidation of carbohydrates with the formation of carbon dioxide and water. In the case of vertebrates the oxygen is transported to the tissue by blood with the help of a respiratory pigment, the *haemoglobin*, but in the case of insects the blood plays very little role in transportation of oxygen to the tissue except the larvae of *Chironomus* which have haemoglobin in blood plasma.

Various types of respiratory organs are present among the insects according to their mode of lives.

1. Tracheal Respiration in terrestrial insects

In the insects, particularly in terrestrial mode of life, the respiratory organs are *tracheal tubes*. The tracheal tubes are invagination of the ectoderm and are internally lined by cuticular intima which is thickened by spiral filaments called *taenidia*. It not only gives the supporrt to tracheal branches but also prevents the collapse of wall. These tracheal trunks divide and redivide and ultimately ends into a cluster of tubes called *tracheoles*. The tracheole devoid of taenidia. The tracheoles are lines by a protein called *trachein*. The ultimate trachioles enter the individual cells of the body, and are partly filled with a fluid called *tracheole fluid* at their distal end. By the reduction in the amount of fluid, more of the surface walls of the tubules in exposed to oxygen and more oxygen is immediately supplied to the tissues.

In certain insects like cicadas, honey bees, butterflies etc the tracheae dilate into *air sacs* of different sizes. The air sacs lack the spiral thickening, allow an increased supply of oxygen and afford a greater breathing capacity of the insect.

The tracheal branches open to outside by many apertures called *spiracles*. The spiracels are elongated or circular apertures and situated on the lateral sides of the body. Usually these are 10 pairs in number, 2 pairs are thoracic and 8 pairs abdominal. Each spiracle is supported by a small annular sclerite called *peritreme*. The spiracles can be closed or opened by spiracular muscles. In some cases the spiracels are absent so that on the basis of presence or absence of spiracels the system is divided into open type or closed type

In the open type of tracheal system when all the spiracles remain open it is called *holopneustic,* as found in most of the adults and some larvae. When on or more pairs of spiracels remain closed the system is called *hemipneustic*. It is seen in some flies, butterflies, beetles etc, when the posterior spiracels are functional, the system is called *metapneustic*. When the functional spiracles are present only on prothorax, it is called *propeneustic* when the spiracels are absent it is called *apneustic*.

II. Close type of respiration in aquatic insects.

The close type is present in some aquatic insects and endoparasitic larvae. Insects living in aquatic medium have different adaptations. Inscets of order Odonata, Placoptera and Ephemeroptera are primarily aquatic where as insects of order Hemipters, Coleoptera, Lepidoptera, Diptera and Neuroptera are secondarily aquatic. Most of the forms are aquatic only during larval stages.

The inmature stages of Odoneta, Plecoptera and Ephemeroptera are aquatic and draw oxygen through general body surface or specialised structures, the *gills,* so that they have two types of respiration:

1. Skin or cutaneous respiration.

2. Gill respiration.

1. *Cutaneous respiration.* In those, insects where cutaneous respiration occurs, the tracheal system is fairly developed without spiracles. The gaseous exchanges takes place through general body surface by diffusion. The larvae of aquatic lepidopterans show this type of respiration.

2. *Gill respiration.* It provides an enlarged respiratory surface formed by evagination of integument. The gills are usually thin walked. In aquatic froms three types of gills are present:

(i) Tracheal gills

(ii) Spiracular gills

(iii) Blood or anal gills.

(i) *Tracheal gills.* It is formed by hallow evagination of integument or intestinal wall. These are supplied by fine tracheal branches and so known as tracheal gills. The tracheal gills may be filamentous (as present in the case of Plecoptera, some Ephemeroptera and Trichoptera) or plate-like (as in larval Odoneta and some Ephemeroptera). In some odonates peculiar *rectal gills* are present. These are six folds formed by the rectal wall. They are well supplied with tracheal branches and hence sometimes they are referred as *branchial basket.* The insect draws water in the rectum through the anus and exchange of gases takes place here.

(ii) *Spiracular gills.* These gills are found in some coleopteran and Dipteran aquatic pupae. Here 1 Or 2 pairs of spiracles are drawn out to long processes which have named as spiracular gills by *Hinton* (1953). Some of the pupae can live in damp climate. So that the gills are considered to adopted both for aeral as well as aquatic respiration through gill surface.

(iii) *Blood or anal gills.* They are thin diverticulae of integument or of the proctodaeum. They are present in aquatic larvae of beetle (*Hydrobia).* These gills are supplies with blood hence they are known as blood gills but in the case of larvae of *Tipulid* the tracheal branches are also present so that they function as the tracheal gills as well as blood gills. In the case of blood gills the respiratory function has not been demonstrated. *Wig-*

glesworth (1953) have shwon in mosquitoes larvae the blood gills are concerned with the regulation of movement of ions abosrbed from the water.

III. Open type of respiration in aquatic insects

This type of system is found in secondary aquatic forms belonging to order Hemiptera and Coleoptera. They have spiracles through which trachail branches draw in atmospheric air. Since the insects live in water they have developed certain adaptations to ensure the supply of air while they are in submerged stage. The various modifications are:

1.*Air stores.* Some of the nymphs and adults of water-boatman and the back-swimmers and also the nymphs of *Spherodema,* certain amount of air is carries by the insect attached to some part of body when the insect dips in water. In the case of water boatman and back swimmer air is carried on ventral surface of body. The air is kept by hairs called *hydrofuge hairs*, which resist penetration of the air filled by water. In adults of some beetles the air is carries by wing cover or elytra. The spiracles remain in contact with air and draws it but the air is not available for a long time so that the insect have to come again and again to take fresh air bubbles to the surface. Besides in respiration the air bubbles also has hydrostatic function.

2.*Gaseous plastron.* This plastron is a permanent bubble held by various means around the body by certain insects. The plastron communicate with spiracles so that the spiracles are able to draw air from water. So that it is permanent nature. The insect has not to come an surface again and again for fresh air. *Torpe* (1950) described such type of respiration in certain aquatic insects.

3. *Air tubes.* Many insects which remains submerged in water acquire and obtain air through one or two tubes which can break the water surface. In such insects the spiracles are situated at the tip of the tube and the rest of spiracles are non-functional and hence either may closed or under-developed (larvae of mosquitoes). In the *Culex* larvae a tube projects above the surface of water to draw air in. The rat-tailed (dipteran larvae) have the tube of about 3-4 inches long. When the larva remains submerged in water the long tube project outside to draw atmospheric air. While this type of tube present in different form in some aquatic

Hemiptera such as *Nepa, Belostoma* and *Spherodema,* their is a long respiratory spihon through which air is drawn.

4. *Plant perforating spiracles.* In some insects the spiracles are situated on sharp - pointed processes which are penetrated by insect into the air cavities of hydrophytes for the purpose of getting air. This is found in the larvae and pupae of certain colepterans and dipterans.

Thus from the above account, it is clear that aquatic insects have assumed several modifications in the respiratory system to enable them to live in water, Without such modifications it would not have possible to them to lead an equatic life.

14

METAMORPHOSIS IN INSECTS

Insects develop from the primordial cells in the ovaries of females. The eggs are laid and hatched out side the body after a variable time. The development within the egg in called *embryonic development* and period after hatching of egg is called *post-embryonic development*. One of the most characteristic features of insects is the fact that they almost always hatch in a condition morphologically different from that of the imago. In order to reach the latter instar they consequently have to pass through changes of form which are collectively termed *metamorrphosis*, these are usually most pronounced towards the end postembryonic development and they are accompnied by physiological and biochemical changes. During metamorphosis the exoskeleton or outer skin is cast of periodically. This process is knwon as moulting or ecdysis. The larval stage between two moultings in called *stadium*. Thus the insect growes by a serics of changes and these growing stages are called *instars*. The number of moults vary but is constant for a given species. On the basis of these changes the metamosphosis is distingusihed into following types:

1. Ametabalous development or no metamorphosis
2. Hemimetabolous development or incomplete metamorphosis
3. Paurometabolous development or gradual metamorphosis
4. Holometabolous development or complete matamorphosis
5. Hypermetabolous development

1. *Ametabolus development or development with no metamorphosis* - In some primitive forms of insects the egg hatches in a form that resembles the adult in all morphological details except in size and colour and in the absence of wings, reproductive organs and external genitalia (styles etc.). The stages of development are : egg-young-Adult. Ametabolus development is seen in *Lepisma*, spring tails *(Thysanura, Collembola and Protura i.e.* in apterous insects).

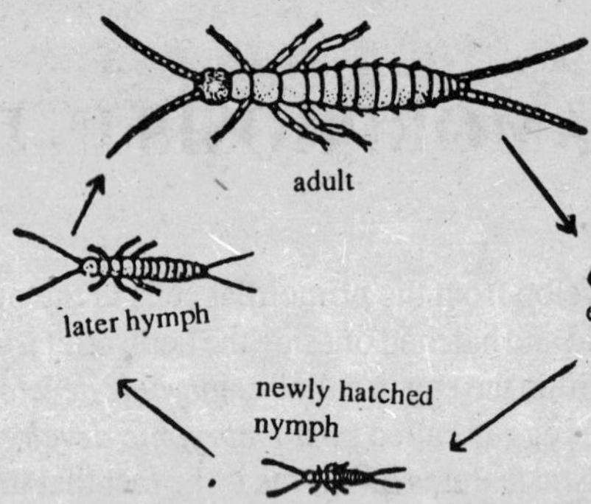

Fig. 14.1 (*Lepisma*).

2. *Hemimetabolous development or development with incomplete metamorphosis*. This type of life-cycle includes gradual metamorphosis as in paurometabolous type except that the eggs are laid in water and

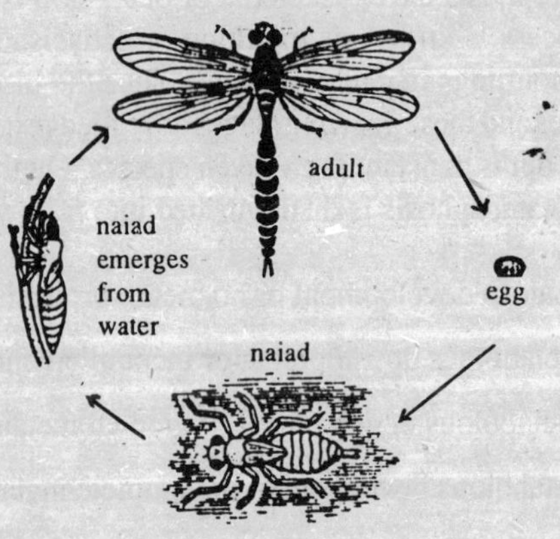

Fig. 14.2 Hemimetabolous development.

develop in aquatic nymphs or naiads. These naiads posses tracheal gills but adults are formed from naiads by repeated moults and maturity. Final moult results in the winged adult. The developmental stages are: Egg-Naiad-Adult. Hemimetabolous development is senn in Mayflies, Dragonflies, Stone flies etc.

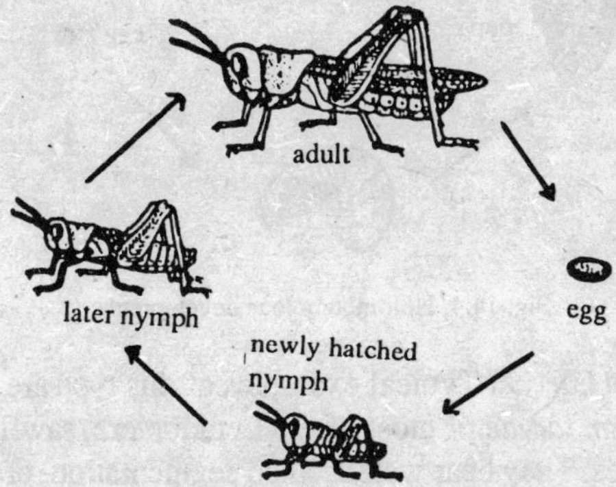

Fig. 14.3. Paurometa bolous development

3. *Paurometabolous development or development with gradual metamorphosis.* In case of gradual metamorphosis, the newly hatched creature resembles an adult in general body form, but lacks wings and external genital appendages. The young or the *nymph* undergoes several nymphal stages through successive moultings to become an adult. The developmental stages are: Egg - Nymph--Adult. Paurometabolous development occurs in grasshopper, cockroaches etc.

4. *Holometabolous development or development with complete metamorphosis* - In complete metamorphosis the development includes four developmental stages - egg, larva, pupa, adult. The larva that hatches out of the egg is totally different from the adult both in structure as well as in habit. The larvae are classified into following types:

(i) *Protopod larvae.* This type is found in the primary larvae of certain parasitic Hymenoptera. The eggs of such species contain little yolk and the larvae emerge is an early embryonic phase. The nervous and respiratory systems are undeveloped.

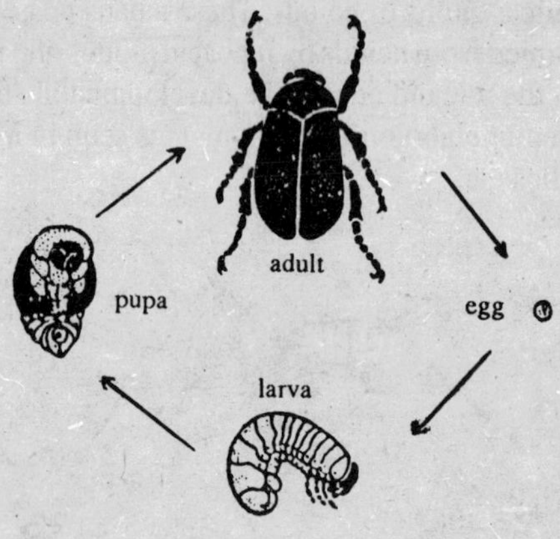

Fig. 14.4. Holometabolous development

(ii) *Polypod Larvae*. Typical examples of this type are the so called *eruciform larvae* of most of the Lepidoptera, sawflies and scorpion flies. They bear well defined segmentation, presence of abdominal limbs or prolegs and a peripneustic tracheal system. Antennae and thoracic legs are small, compound eyes, ocelli and wing pads are totally absent.

(iii) *Oligopod Larvae*. These are characterized by the presence of more or less well-developed thoracic legs and absence of abdominal appendages except sometimes for a pair of similar caudal process. Two common types can be distinguished:

(a) *Campodeiform larva*. They are so called from their resemblance to *Campodea* (Thysanuran insect). It bears an elongated more or less fusiform somewhat depressed body which is often well sclerotized and markedly prognathous head, long thoracic legs and usually a pair of terminal adbominal process. Compound eyes, dorsal ocelli and wing pads are absent. Campodeiform larvae occur in the Neuroptera, some Coleptera, Trichoptera and Strepsiptera.

(b) *Scarabeaeiform larva*. These are subcylindrical, C- shaped larvae with shorter thoracic legs, a soft fleshy body and no caudal process. The larvae are seen in Scarabaeoidea, in Coleopteran familes etc.

(iv) *Apodous larvae*. In this type the trunk appendages are completely absent and most cases are probably derived from oligopod type.

Such a larvae are seen is several familes of Coleoptera.

After several changes the larva turns into pupa. The pupa appears to be inactive externally but internally it is very active and forms the imago. The following types of pupae are distinguished by *Hinton* (1946):

(a) *Decticous pupae*. These have relatively powerful, selerotized, articulatede mandibles. This is the primitive type of pupa and is always of exarate i.e. the appendage, are not closely applied to the body but are used in locomotion. This type of pupa occurs in Neuroptera, Macoptera, most Trichoptera, in some lapidopteran families etc.

(b) *Adecticous pupae*. This type have a non-articulate mandibloes and are not used in escaping from the cocoon. Three forms of adecticous pupae are recognizable.

(i) *Exarate a decticous Pupae*. The appendages are free of any secondary attachment to the body. They are found in the Siphonoptera and Strepsiptera, most Coleoptera and Hymenoptera.

(ii) *Obtect adecticous pupae*. Here the appendages are firmly pressed against its body and are soldered down to it by a secretion produced at the last larval moult. These pupae occur in all higher Lepidoptera, some Coleoptera etc.

(iii) *Coarctate Pupae*. This is an adecticous as well as exarate type of pupa which is enclosed in puparium. The puparium is formed from preceeding larval cuticle. It does not show any changes in external appearence. It is seen in Cyclorrhaphan Diptera.

5.*Hypermetabolous development or development with hypermetamorphosis*. When a developing insect passes through two or more markedly different larval instars it is said to undergo *hypermetamorphosis*. The last or final larval instar changes into non feeding, inactive stage called *prepupa*. It is enclosed in the cuticle of preceeding larval stage. Such a metamorphosis occurs in some Neuroptra, Coccidals and many endoparasitic Hymenoptera.

Internal Metamorphosis

In those cases where the metamorphosis is gradual or incomplete, the internal organs do not undergo any major modifications but some very minor changes are observed. In the case of holometabolous and hypermetabolous development major changes take place in internal organs because of different mode of life. All these modifications of interanl organs takes place within the pupa. There may be other structural or physiological internal changes, so that the pupa is inactive externally but it is active internally. During early pupal stages internal organs break down or dissolved (probably as a result of lysosomal breakdown) or by *phagocytes*. This process in known as *histolysis*. This phenemenon differs in different insects. In some cases it is minor or may be major. After histolysis new tissues are formed from certain pad-like structures known as *immaginal pads or histoblasts*. This process is known as *histogenesis*.It is ultimately followed by a process of *organogenesis*. The internal changes within the pupa differ in different groups of insects. In coleoptera, Lepidoptera, Hymenoptera and some Diptera intestine and other larval parts are greatly modified without general histolysis. Similarly in certain dipterans the general histolysis is noted so that all the organs are discolved and reformed except the nervous system, heart, reproductive organs and thoracic appendages. The study of internal metamorphosis is not thoroughly studied in all the groups of insects but still it is unknown in some of the other groups of insects.

Hormones and their role in metamorphosis

Insects have a rigid exoskeleton, because the exoskelton do not expand to allow for enlargement, insect growth is associated with the periodic shedding of the exoskeleton in a process known as *ecdysis*. Ecdysis is followed by the expansion and hardening of a new, soft, larger exoskeleton that has developed underneath the old exoskeleton. Thus, insect development includes a series of ecdysis or moults, and the stages between the moults, are known as *instars*.

In insect development, *metamorphosis* is the change in body from that converts an immature individual into an adult. Metamorphosis is either incomplete or complete comprising eggs-nymphs and adults in former case and eggs - larvae - pupae - adults in latter case. Several hormones regulate insect growth, moulting and metamorphosis. The neurosecretory cells of the brain produce a *brain hormone* that travels along axons to a structure called *corpus cadiacum,* which is attached to the brain. From the corpus cardiacum, the brain hormone is released and carried to the *prothoracic glands*, which is stimulated to release *ecdysone*.

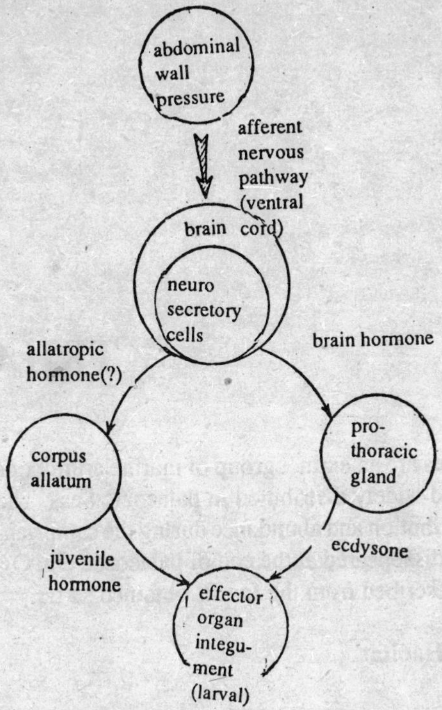

Fig. 14.5. Control of moult and metamorphosis in insects.

Ecdysone is also called the moulting hormone or moult and maturation hormone because it promotes moulting and tend to spur the developmental process along toward pupation and adult hood. During larval stage, the effect of ecdysone is countered by the action of *juvenile hormone*, secreted by corpus allatum. When Juvenile hoermone is secreted, the organism remains immature. In the last larval stage, however, juvenile hormone secretion decreases dramatically and the animal pupates and begins its transformation to the adult form.

From an evolutionary point of view, it is intersting that some plants produce compounds similar or identical to ecdysone or juvenile hormones. This apparently protects the plants by disrupting the development of insects that eat them.

15

TRILOBITES

Trilobites are an extinct group of marine arthropods that were once abundant and widely distributed in palaeozic Seas. They reached their hight of distribution and abundance during the camprian and Ordovician periods and disappeared at the end of palaeozoc era. Over 3900 species have been described from the fossils obtained so far.

Habits and Habitat

The trilobites were exclusively marine since they were found only in association with the remains of marine fossils. They were mostly bottom dweller, crawling over sand and mud with the help of legs. They probably swallowed mud and slit and used the organic part as food, like many annelids.

External features

Size. Most trilobite species ranged from 3-10 cm in length, although some planktonic forms were only 0.5 mm in length. Some species attained a length of 67.5 cm (*Terataspis).*

Division of Body

The body was divided into three more or less equal sections:

1. Cephalon or Head

2. Thorax or Trunk
3. Pygidium.

Each of these body division was in turn divided into three regions by a pair of furrows running from anterior to posterior and forming a medians *axial lobe* flanked on each side by a *lateral* or *pleural lobes*. In some cases the thorax and abdomen formed the *opisthosoma*.

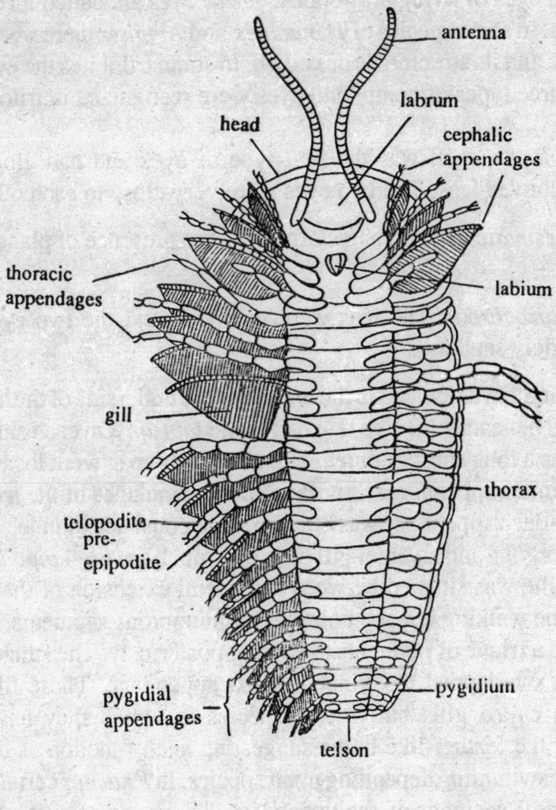

Fig. 15.1. *Triarthrus eatoni* (

1. *Head or cephalon.* It was distinct, solid, unsegmented, semicircular and often prolonged backwards at the angles. The head was covered by a carapace or*cephalic shield* with the posterior lateral margins

projecting backward. The head was composed of atleast five fused segments. The 5th *occipital segment* was most distinct. The original head segmentation was frequently indicated by transverse grooves on the carapace. The pleural skeleton was reduced to thin membranes to which the appendages were attached. The lateral portions of the carapace were known as *cheeks or genae*. The check of either side was traversed by an oblique *frontal, facial* or *ocular suture*. On each side of the middle of the carapace was a pair of eyes that varies considerably in size, depending upon the species. X - ray studies indicate a structure similar to the compound eyes of living arthropods. These eyes are called maculae by Lindstrom. In some species of *Trinuclues* and *Amphyx* there was a single simple eye and its structure is unknown. In some trilobites the eyes were absent. Three types of compound eyes were recognized in trilobites:

(i). *Holochroal*. It was the first type of eyes and had globular or biconvex lens. The two eyes were very close to each other.

(ii) *Presmatic*. It was characterized by the presençe of planoconvex lens.

(iii) *Schizochroal*. The lens were biconvex and the two eyes were widely separated.

The mouth was located in the middle of the underside of the head just behind and beneath a lip-like flap called the *labrum*. On each side of the labrum was a long sensory *antenna*. Behind antennae were located four pairs of similar appendages, similar to the appendages of the trunk and pygidium. Each appendage was biramous and consisted of inner walking leg or *telopodite* and an outer gill bearing branch are *pre-epipodite*. The pre-epipodite was situated between the lateral extension of the pleural lobe and the walking leg and consisted of numerous segments, each of which bore a fringe of filaments extending posteriosly. The fringe of one appendage overlapped the fringe of next appendage. These filaments have been called gills but in few species in which they have been preserved, the feather-like barbs suggesting such function as digging, filtering or swimming depending upon species. In *Phacops* certain of the pairs of basal segment of the appendage i.e. coxae carried opposing toothed processes that must have functioned as *jaws*.

2. *Trunk or Thorax*. The trunk consisted of varing number of separately articulating segments. The number may be 2 in *Agnostus* to 29. Owing to the mobality of thoracic segment some trilobites were found

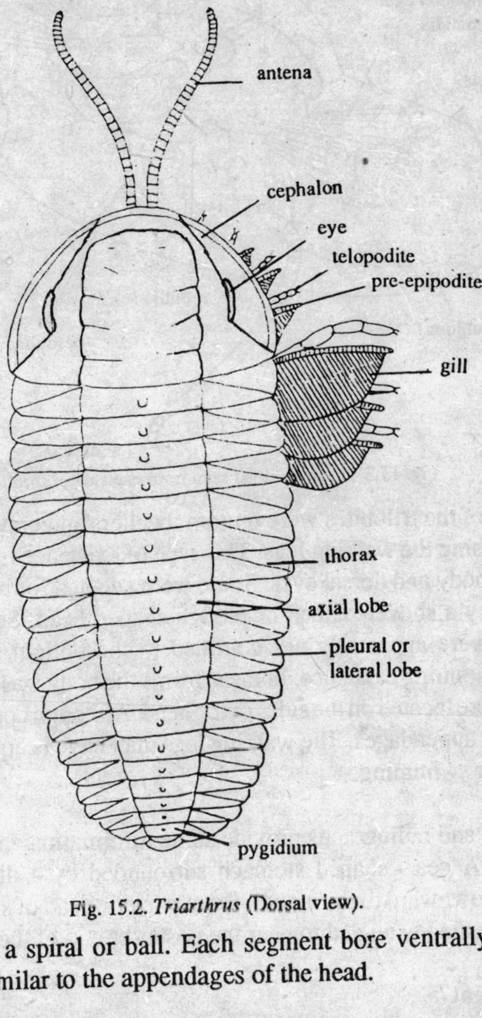

Fig. 15.2. *Triarthrus* (Dorsal view).

rolled up into a spiral or ball. Each segment bore ventrally a pair of appendages similar to the appendages of the head.

3. *Pygidium.* The pygidium was constricted on the same plan as the thorax except that the segments were fused and formed a solid shield. The appendages of pygidium were sucessively smaller towards the posterior end. The number of abdominal segments varies from 2-28 and the last segment of pygidium was called *telson*. The *anus* was situated on underneath of the telson.

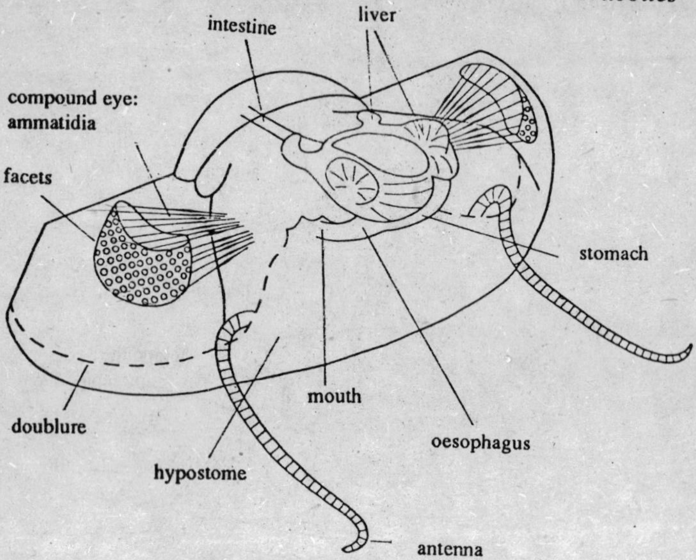

Fig. 15.3 Trilobite. Head seen from one side of front.

Most of the trilobites were bottom dwellers and crawled over sand and mud using the walking legs. This type of existence is proved by the flatenned body and dorsal eyes. Some were adapted for burrowing as is evidence by a shovel-shaped or plough-shaped head. Some groups of trilobites were apparently not confined to the bottom but took up a pelagic swimming existence. In these forms the body was narrower, and the eyes were located on the sides of the head. Although nothing is known about their appendages. The walking legs may have been flatenned and adapted for swimming.

Molds and radiographs provide some information about the digestive tract. A pear- shaped stomach surrounded by a digestive gland extended downward to the mouth. From posterior end of stomach a long narrow intestine extended to near the posterior end of the body.

Development

The larvae of several species of trilobites have been found in the fossil state. For some forms, such as species of *Sao* and *Olenus*, almost the entire larval series has been determined from fossil specimens. During their postembryological development, trilobites passed through three larval periods, each of which consisted a number of instars. The first larval stage was *protaspis larva*. It was a small planktonic form measuring from 0.5-1 mm in length. The body had a single dorsal carapace consisting of the acrom and four postoral segments. After passing through several instars, in which additional segments were added to the carapace the protaspis larva passed into a *meraspis larva*. In this larva the pygidium was located behind the cephalon. The thoracic region appeared during succeeding molts as segments were gradually freed from the anterior border of the pygydium. The final larval stage was *holaspis larva,* though still very small, contained the definite number of adult thoracic segments. Nothing is known as larval appendages and none of the stages discovered so far.

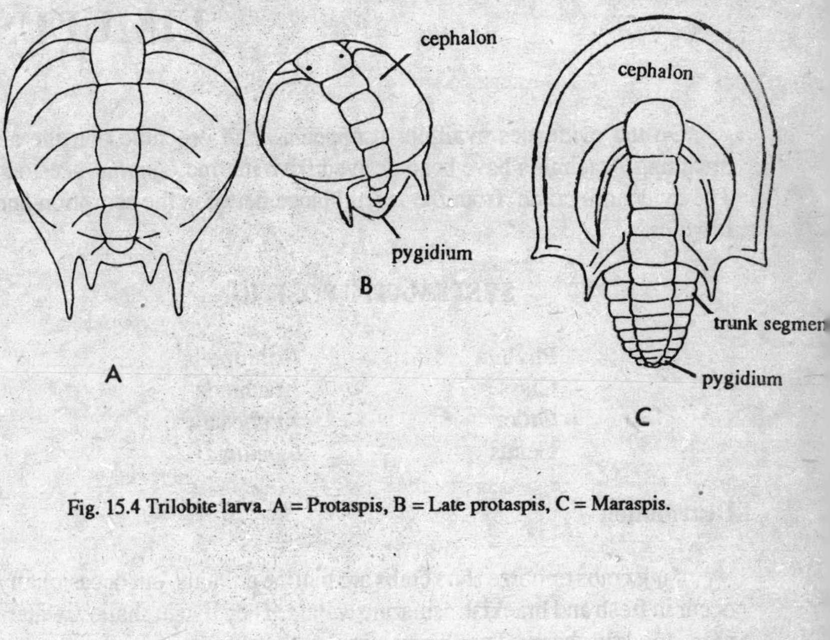

Fig. 15.4 Trilobite larva. A = Protaspis, B = Late protaspis, C = Maraspis.

16

LIMULUS

On the evidences availble it appears quite probable that the air breathing Arachnida have been derived from marine *Limulus* acestors. The evidences come from the resemblance between the scorption and *Limulus*.

SYSTEMATIC POSITION

Phylum	-	Arthropoda
Class	-	Arachnida
Order	-	Xi phosura
Genus	-	*Limulus*

Distribution

King crabs or horse shoe crabs are marine animals, but occasionally occur in fresh and braekish estuarine waters. They live in shallow waters along muddly shores. They are noctural as they become active and swim through the water at night.

External features

Shape and size. The body is depressed, dark brown in colour and measuring about 60 cm in length. They are called kingcrabs but neither they are kings nor crabs.

Division of Body. The body is divisible into *cephalothorax, abdomen* and a *caudal* spine or *telson*. The body is covered by smooth chitin. The cephalothroax is semi-circular and consists of a pair of *median eyes* which are very close to each other. There is a pair of lateral eyes which are large and widely separated from each other. The median eyes lie on median ridges whereas the lateral eyes on lateral ridges. Seven spines are also present. On the ventral side of cephalothorax there is a *mouth* and seven pairs of appendages. Infront of mouth there is a flatenned longitudinal ridge called *camerostome* and beyond the mouth is an oval structure

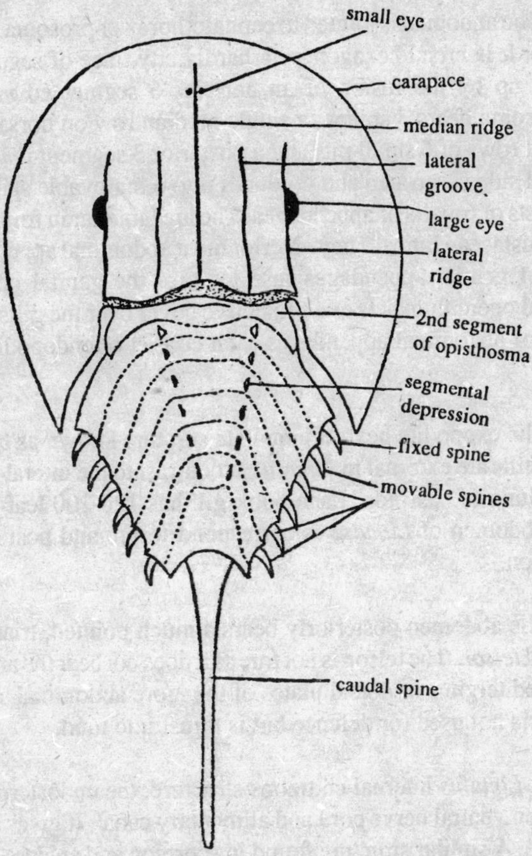

Fig. 16.1. Dorsal view

called *promesosternite*. The mouth is surrounded by seven pairs of appendages. The first pair is called *chalicerae*. Each chelacera is a 3 segmented structure which is followed by 5 pairs of walking legs called *pedipalpi*. Each pedipilp is made up of six segments. In female the fourth anterior pair of *pedipalpi* are chelate in appearence whereas in the males the first pair have claws which are curves. The 5th pair in male differ from the female by possession of a peculiar curved spine called *spathulate* process. In scorpion cephalothorax bears 6 pairs of appendages with a pair of median eyes and many lateral eyes. All the appendages are postoral in position but at a later stages the first pair become preoral in position.

The abdomen is jointed to cephalothorax or prosoma by a transverse hinge. It is broad hexagon with hardly any trace of segmentation. It is made up by the fusion of an anterior 6 segmented *mesosoma*. The mesosoma bears 3 spines in single median row on dorsal side and two lateral rows of 6 small pits and a posterior 3 segmentd *metasoma* .Each lateral side is serrated and produces into 6 inmovable spines. Abdomen consists of 6-pairs of appendages. The are lamellar in form and resemble the crustacean limbs. They anterior most abdominal appendages the first pair of gential appendages fuse to form the genital operculum. The genital operculum acts as a protective cover over the gills. Among the 5 pairs of abdominal appendages each consist of endopodite and exopodite.

The exopodite bears a lamellate surcture known as *book gills*. The book gills are external in *Limulus* in contrast to the interal appearence of book lung of scorpion. Each book-gill has 120-200 leaf-like lamellae. The abdomen of *Limulus* is correspond to pre-and post abdominae of scorpion.

The abdomen posteriorly bears a much pointed, triangular, spear-shaped *telson*. The telson is not true as it does not bear the anus. It consists of fused tergites or dorsal plates of the more abdominal segments. The telson is not used for defence but is thrust into mud.

In *Limulus* internal chitinous structure, the endosternite present in between ventral nerve cord and alimentary canal. It gives attachment to muscle. A similar structure found in scorpion and spiders besides a few crustacean like *Apus*.

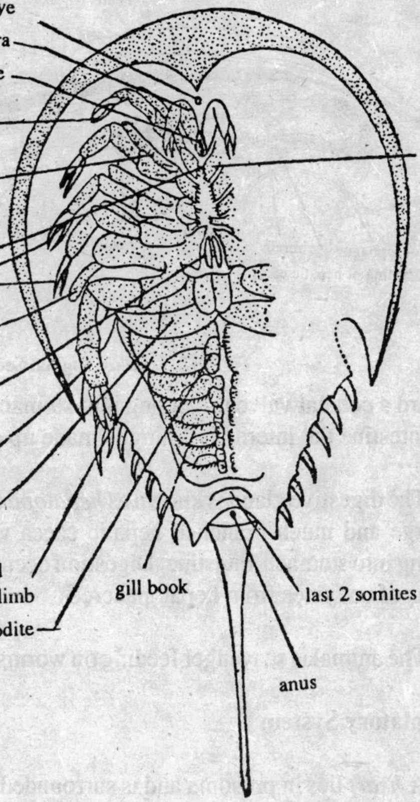

Fig. 16.2. *Limulus*.. Ventral view

Digestive System

The *mouth* is ventrally placed and followed by *oesophegus*. It passes upward from the mouth, through the vascular ring, then runs forwards to the more swollen *crop*. This curves upwards and backwards, merging into the *gizzard*. Gizzard is followed by *stomach - intestima*. A short *rectum* leads to the *anus*.

The lining of oesophagus is longitudinally folded. Lining of crop is wrinkled and the gizzard is much less distensible, and its lining is thrown into longitudinal ridges studded with somewhat flatenned teeth. From

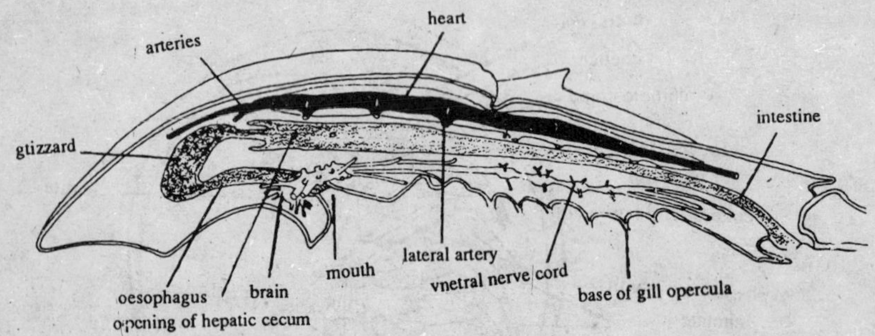

Fig. 16.3. *Limulus*. Sagittal Section

gizzard a conical valve projecting into stomach-intestine. Except stomach-intestine the internally lining is made up of cuticle.

The digestive gland is known as *hepatopancreas*. It consists of a pair of large and much branched hepatic caeca which remify. Two ducts leading into stomach-intestine. Digestion occurs in stomach-intestine by means of enzymes from hepatopancrees.

The animal is scavenger feeding on worms, soft mollusks and algae.

Circulatory System

A *heart* lies in prosoma and is surrounded by a thin-walled pericardial sinus. It is suspended from the walls of pericardial sinus by 9 pairs of elastic ligaments . The heart is tube-like and provided with 8 pairs of *ostia*.

The arteries leave the heart are four pairs of *lateral arteries*, one pair of *aortae* and a *medium frontal artery*. Among them the first pair is known as *aortic arch* and fuse to form *vascular* ring. The ring gives rise to small vessels which supplies the walking legs and other median and internal structures. The artery arising from the vascular ring extends posteriorly and gives rise smaller vessels in gills. The blood from the median artery is collected by *supra anal vessel* which bifurcate to give rise thecollateral vessels. *Colleteral vessel* communicates the heart by 4-pairs of lateral vessels. The blood from the gills is collected back by longitudinal vessel.

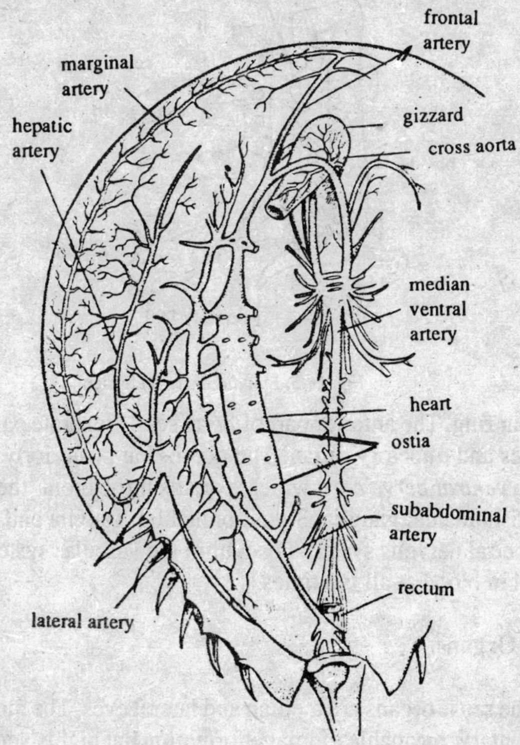

Fig. 16.4. *Limulus*. Arterial System.

The colour of blood is bluish due to the presence *haemocynin*. Amoeboid cells are also present in blood.

Excretory System

There is a pair of *coxal glands* which are brick-red in colour. Each gland consists of four lobes and inside each lobe there is a network of tubules. At the base of 5th appendage the two lobes of a side open by means of a duct.

Nervous System

It is highly concentrated and consists of two pairs of ganglia or *brain* united by commissures. The commissures form a ring which lies within

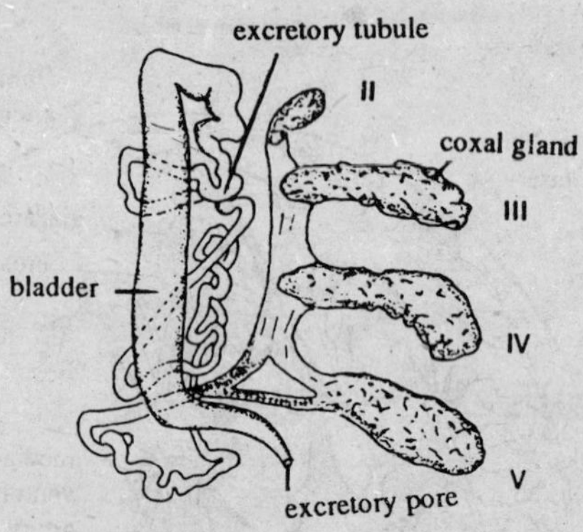

Fig. 16.5. *Limulus*. Arterial System.

vascular ring. The anterior part of brian sends separate pairs of nerves to the eyes and olfactory organ. The nerve ring posteriorly gives rise to a double *ventral nerve cord* which extends throughout the abdomen and bears 5-branchial ganglia, 3 post-branchial ganglia and lateral nerves. The central nervous system lies within the vascular system, so that it is dipped in blood at all the times.

Sense Organs.

The sense organs are median and lateral eyes. The median ocelli are rudimentary, incapable of image formation but highly sensitive to ultraviolet light and are analogous to the pineal body of vertebrates. The lateral eyes bear a number of widely spaced *ommatidia* beneath a single lens. Each ommatidium consists of central ganglion cell surrounded by 9-11 *retinular cells*. The retinulae and ganglion is formed by hypodermis. The lens is cuticular in nature. The image formed is of compound nature.

Reproductive System

The sexes are separate. Males are smaller than the females and their first walking legs are non-chelate but clawed, serving as claspers during pairing.

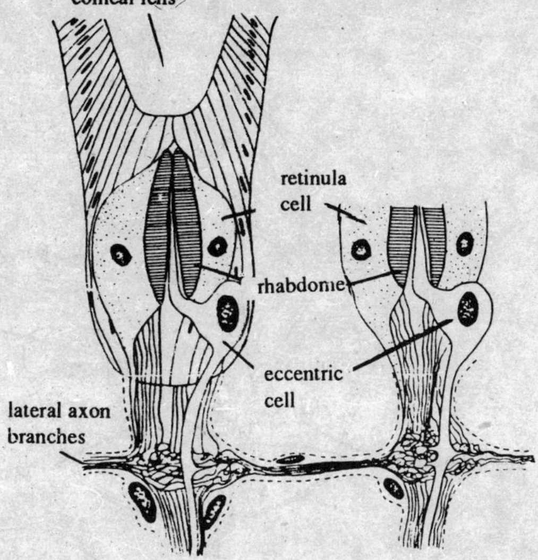

Fig. 16.6. *Limulus*. L.S. of eye through two ommatidia.

In males the *testis* consists of an intricate reticulum of tubes with small follicles on the wall of many of them. A globular *sperm sacs* are present in which early stages of spermatogenesis occur. *Vasa deferentia* are located in the hepatopancreas. Each *spermatozoon* is provided with a long, vibratilc tail.

In female the *ovary* is very similar in structure and extent to the testis of the male. The *oviducts* are large, muscular tube, each of which starts medical and posterior to the block of muscular supplying the 6th appendage and passes posteroventrally to its opening on the genital operculum.

The male genital opening on the operculum are round and each is located on a well-defined papilla. In female each opening is a transverse slit and there is no definite papilla.

The breeding season is spring and early summer. For mating and egg laying the males and females migrate into shallow water. Spawning occurs at night, when male climbs onto the abdominal carapace of the

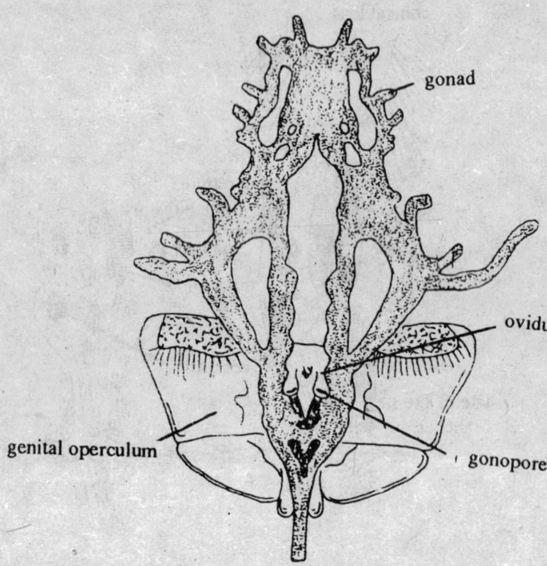

Fig. 16.7. *Limulus*. Female reproduction organ.

female. Meanwhile, the female scops out a series of depressions in the sand and deposits 200-300 large eggs. The eggs in each depression are fertilized by the male during their deposition. The eggs are centrolecithal and are covered by a thick chorion. Cleavage is holoblastic. A *trilobite* larva emerges from the egg. It actively swims about and burrows in sand. The little caudal spine does not project beyond the abdomen. Only 2 pairs of book gills are present instead of 5 pairs. With the successive moults the remaining book gills appear. The caudal spine increases in length. Sexual maturity is not reached until the third year. In Japanese species attainment of sexual matirty requires 13 moults in male and 14 moults in female.

Affinities of Limulus

I Affinities with Arachnids

Lankester and associates have compared scorpion and *Limulus* in very much details and established a close relationship between the two. The points of resemblance between the two are:

1. The cephalothorax bears 6 pairs of appendages.
2. The appendages are postoral but for the first becomes preoral in position during development.
3. Chilaria of *Limulus* is respresented by a pentagonal plate. In scorpion it lies infront of operculum.
4. Both bear median and lateral eyes.
5. Abdomen of *Limulus* corresponds with the pre and post abdomen of scorpion.
6. In both the first pair of abdominal appendages forms genital operculum.
7. The second pair of abdominal appendages is modified into pecten in the scorpion and rest of the appendages disappear but in *Limulus* all the appendages are present and bear book-gills.
8. Structure of book gill comparable to the book lungs of scorpion.
9. Hepatopancreas is present in both with reproductive organs embeded.
10. Coxal gland (one pair) is present in both.
11. There is a similarties between the development of *Limulus* and scorpion.

II. Affinities with Trilobita

Lankester put Limulus and Trilobita in the same phylum. The larva of *Limulus* shows superficial resemblance to a fossil Trilobite.

1. Body is trilobed divided by a pair of longitudinal grooves into a median and two lateral lobes.
2. Abdomen distinctly segmented.
3. Presence of lateral eyes.

It, however, differs from Trilobita in the absence of antennae and the appendages show marked differences. In Trilobita the appendages show marked differences. In Trilobita the appendages are similar and biramous.

III Affinities with Eurypterida

Eurypterids are extinct relatives of *Limulus*. The chief resemblance between two are:

1. Both are aquatic.
2. Respiration by gills.
3. Carapace dorsally bears paired median simple and lateral compound eyes.
4. Position of appendages similar.

It seems that marine *Limulus* ancestors gave rise to the terrestrial air breathing animals like scorpion. During the change of aqatic to a terrestrial life change took place in respiratory system. So that book-gills modified into book lungs. The structure of book-gill and book-lung is, however, similar though they are differently situated. The book-gills are external in position and lie on the external abdominal appendages in *Limulus* whereas the book-lungs are situated inside the pulmorary sac or pits. These pulmonary sacs are necessary for terrestrial mode of life. Probably the tracheae of other arachnid arose as modification from the pulmonary sac. The evidence that the trachea developed from pulmonary sac come from the fact that certain terrestrial isopods the tracheae develop from endopodite. Therefore, it seems proper that *Limulus* like ancestor which were marine gave rise to air breathing arachnids like scorpions and later give rise to other tracheal air breathing arachnids, probably by modification of pulmonary sac into trachea.

17

SCORPIONS

The scorpions are the oldest known terrestrial arthropods and they have been the first members of this phylum to have conquered land. The scorpions are most common in tropical and subtropical areas except New Zealand and oceanic islands.,They are most abundent in the Gulf states and Southwest. About 800 species have been described some common genera are *Palamneus, Buthus* etc.

SYSTEMATIC POSITION

Phylum	-	Arthropoda
Class	-	Arachnida
Order	-	Scorpionda
Type	-	Scorpion

Habits and Habitat

Scorpions are generally secretive and nocturnal, hiding by day under stones and wood and in burrows in the grounds. But there are species associated with vegetation. They are often found near dwellings. Scorpions are popularly believed to inhibit desert regions, but although many desert species exist, they are by no means restricted to arid regions. Many scorpion species require a humid environment and live in tropical rain forests and similar jungle habitats. They are carnivorous and feed on small arthropods, especially spiders and nocturnal insects, cannibalism is also practised. These are solitary animals and never live in groups. All the scorpions are viviparous and the females carry the young ones on their backs for some times, about a week.

External Features

Shape and Size. The body is long, narrow and dorso-ventrally flattened. The size varies from 3 to 8 cm. in length. The smallest species is the Middle Eastern *Microbuthus pusillus* wqhich is about 13 mm. long. The largest one is African *Pandinus imperator* which measures about 18 cm. in length. The largest Indian species, *Palamnaeus swammerdani,* grows to a length of about 15 cm. while *Buthus tamulus,* a common species, reaches a length of nearly 8 cm.

Colouration. The colour is variable, usually corresponding with the habitat of the animal which is generally blackish dorsally and slightly light coloured ventrally. The species, which are living in tropical jungles, are of shining black colour, and the species which are found in the sand are pale-yellow in colour usually the dorsal surface is much darker in colour than those of the ventral surface.

Division of body. The scorpion body consists of a *prosoma* or *cephalothorax* covered by a single carapace and a long *abdomen* or *opisthosoma.*

(i) *Prosoma.* It is the broad flat, anterior region composed of the head and thorax which are fused together. It is formed by the fusion of a *preoral* and *six post-oral* segments but only six are distinguished in the adult. The segments are marked by the presence of six pairs of appendages on the ventral side. Dorsally, the prosoma is covered by a large, shield like exoskeletal plate, somewhat squarish in form. It is known as *carapace*. Its anterior margin is divided into *right* and *left frontal lobes* by a median notch. Two to five pairs of *lateral eyes* and a pair of *median eyes* can be distinguished along the antero-lateral margin and in the middle of the carapace. These are simple eyes. On the ventral surface there is a narrow triangular plate, the *sternum,* situated medially between the coxae of 3rd and 4th pairs of legs.

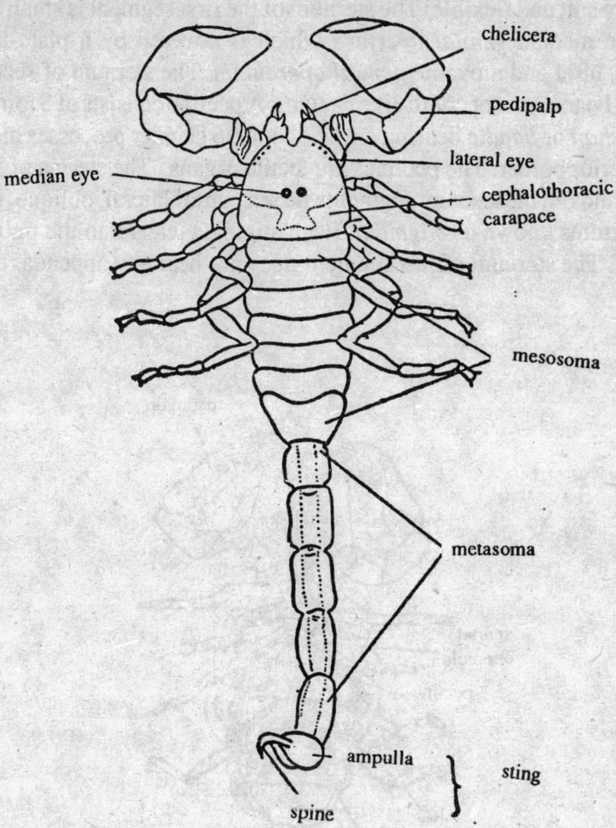

Fig. 17.1. Scorpion. Dorsal view.

A small *mouth* is situated on the anterior end of the prosoma towards the ventral side between the cephalic appendages.

(ii) *Opisthosoma.* The prosoma is followed by a long opisthoma. It is divisible into two parts: anterior *mesosoma or preabdomen* and posterior *metasoma* or post abdomen.

(a) *Mesosoma.* It consists of seven segments in the adult. Each segment of the mesosoma is covered by two firm chitinous sclerites *tergum* on dorsal side and *sternum* on ventral side. The tergum and sternum of each segment are laterally connected by pleural membranes which are soft and flexible. The sternum of the first segment is small and bears the median *genital aperture* which is covered by a plate-like, rounded, bifid and movable *genital operculum.* The sternum of second segment bears a pair of comb-like *pectnes.* A pectine consists of 3 jointed *stem* or *shaft* or *handle* bearing a row of 4 to 36 slender processes along the posterior border. The pectines are tectile organs . The sterma of 3rd, 4th, 5th and 6th mesosomal segments bear a pair of lateral, oblique, slit-like apertures known as *stigmata.* Each stigmata leads into the pulmonary sac. The sternum of last segment does not bear any appendage.

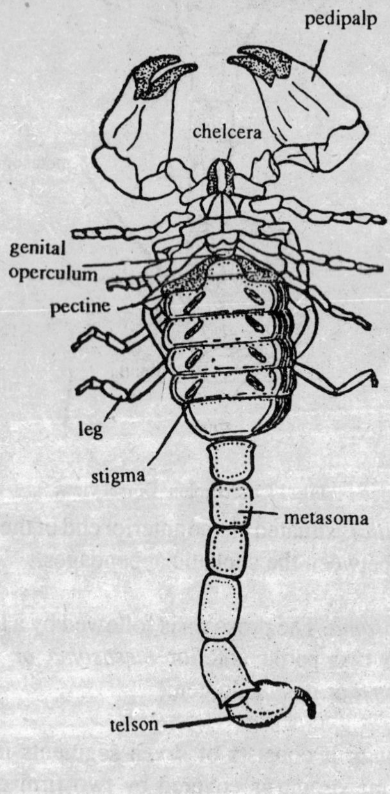

Fig. 17.2 Scorpion. Ventral view.

(b)*Metasoma*. The slender *metasoma* or *post-abdomen* is often wrongly called a tail. It is held upraised, arching over the back in a moving scorpion. It consists of five cylindrical segments, each enclosed with in a complete chitinous ring. The last segment bears the anus on its post-ventral side and a stinging apparatus or *telson*. The telson consists of a swollen base, the *vesicle* or *ampulla*, and a curved and pointed spine, the *aculeus*. It remains attached to the last segment by means of a flexible membrane. Inside the vesicle lies a pair of *poison glands*, the ducts of which open by a pair of minute apertures at the tip of the spine.

The venom of most scorpions, although sufficiently toxic to kill many invetebrates, is not harmful to man. At most, the sting in equivalent to that of a hornet. However, certain species exist that possess a highly toxic venom that can be fatal to man. *Androctonus australis* of the Sahara Desert has venom equivalent in toxicity to cobra venom, and this venom can kill a dog in seven minutes. The neurotoxic venom of scorpins is very painful and may cause paralysis of respiratory muscles or cardiac failure in fatal cases.

Appendages

The cephalothorax bears 6 pairs of appendages, these are a pair of *chelicerae*, a pair of *pedipalpi* and four pairs of walking legs. Of these the chelicerae are preoral and remaining appendages are postoral.

(i) The *chelicerae* are short but stout appendages. They lie close together at the anterior end of the cephalothorax. Each consists of three *podomeres*. The basal podomere is ring-like and concealed beneath the carapace. The middle and distal podomeres together form a grasping apparatus called the *chela or pincer* for holding the prey during feeding. The middle podomere is large, bears setae and is produced on the inner side into a conical toothed process that forms the *immovable finger* of the chela. The distal of interal segment is small, curved and toothed. It forms the *movable finger* of the chela.

(ii) Behind the chelicerae are a pair of powerful and clawed appendages called *pedipalpi*. Each pedipalp is made of six segments, viz, *coxa, trochanter, humerus, brachium, manus* and *movable finger*. The coxa is small and situated on the side of the pre-oral cavity beneath the carapace. On each coxa there is a gnathobase towards the mouth. The gnathobases of both sides serve as jaws, protruding in the preoral cavity and help in squeezing and crushing the body of the prey.

The *trochanter* follows coxa and is irregularly shaped having spinous anterior margin. At the right angle to longitudinal axis of trochanter, there is a stout and powerful *humerus*. The *branchium* lies above the humerus and is directed forward and is provide with granular crests. The *manus* is longest segment and it forms a powerful chela with the movable finger which is movably articulated in its help in siezing the prey.

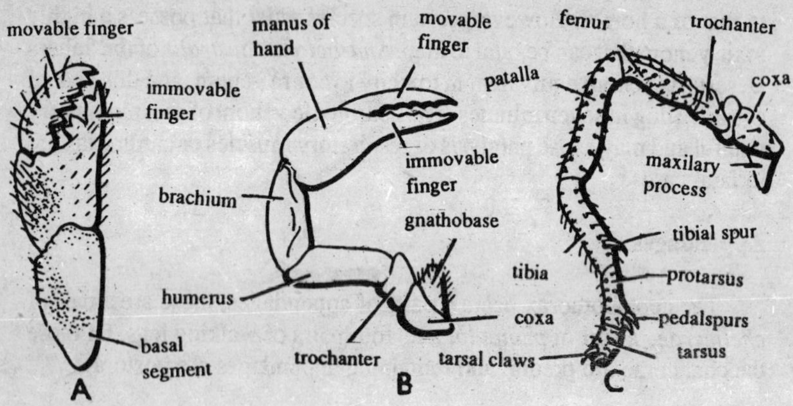

Fig. 17.3. Scorpion appendages A = Chilicera, B= Pedipalp, C = First Walking leg.

(iii) *Walking legs*. There are four pairs of walking legs, attached to the cephalothorax. All the legs are alike and are used for walking. Each leg consists of seven podomeres, *coxa, trochanter, femur, patella, tibia, prototarsus* and *tarsus*.

The coxae of 1st and 2nd pairs are provided with forwardly-directed triangular maxillary processes or gnathobases or endites. Coxae are very small, short, movable, except those of the fourth leg, which are longer. The coxae of the second pair meet with each other in the middle line while the coxae of the 3rd and 4th pairs are immovable and joined together but they do not meet in the middle line of the body and are separated by sternum of the cephalothorax. The *trochanter* is a short and stout segment, moving in all the directions. It is followed by a long, movable *femur*. The *patella* is an important podomere, which gives the leg a double knee, which moves down-wards on the end of the femur. *Tibia* is a part of the leg, which turns down abruptly on the patella. The tarsus is a long subterminal segment. In-between the tibia and tarsus is a structure, called *tibial spur*, while the tarsus bears a pedal spur. *Pretarsus* is provided on its lower side with hairs or spines or both. Each tarsus bears at its apex a pair of large claws, called superior claws and smaller inferior claws. Inferior claws are generally injured and their ends are removed.

Body Wall

The body wall consists of three layers:

(i) Cuticle
(ii) Hypodermis
(iii) Basement membrane.

(i) *Cuticle*. The outermost layer is a brillantly coloures non-living chitinous layer which is secreted by the hypodermis. The cuticle is made up of three layers. (a) a superficial cuticle, *tetostracum* (b) the middle layer, *epiostracum* and (c) the lower lamellous layer, *hypostracum*. The cutcile is traversed by canals which open outside on the surface. In these openings are lodged the duct of secretory glands which are located in the hypodermis.

(ii) *Hypodermis*. It consists of a single layer of columnar cells filled with pigment granules.

(iii) *Basement membrane*. It is a thin and structureless layer. The hypodrmal cells rest on this membrane.

Various kinds of projections are found on the cuticle:

(a) *Dermatidia*. These are small chitinous protuberences on the chela.

(b) *Coelodermatidia*. They occur in the form of hollow projections on the integument. They form hairs, spines, bristles, spur and claws etc. of tarsus.

(c) *Dermal glands.* These are known as *alveolar glands* occur in the chelae, pedipalpi and telson.

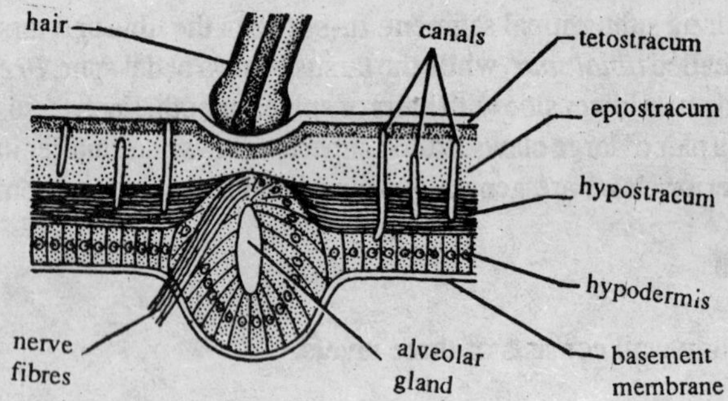

Fig. 17.5. Scopion. T.S. of poison vesicles with two posion of gland and glandula plicata.

Glandula Plicata. The males of some species of genus *Bothriurus* are provides, on the dorsal surface of the poison vesicle, with a scutelliform depression. The chiton of this place is very thick and underneath, the hypodermis forms a special organ consisting of numerous longitudinal folds of cuboidal epithelium. These folds are longer in middle and lower at margins and are glandular. As they are found only in males they are supposed to play some role in sexual reproduction.

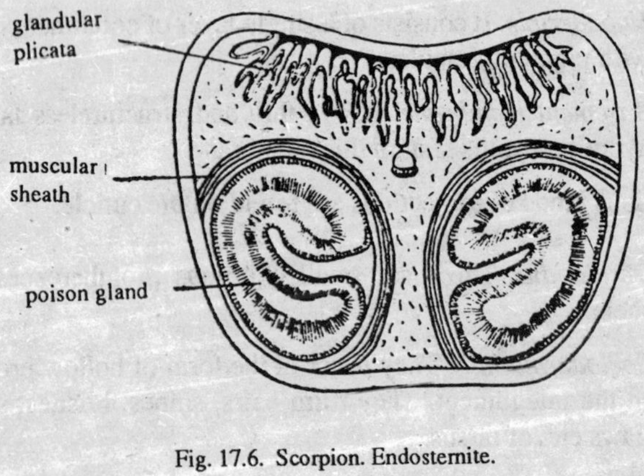

Fig. 17.6. Scorpion. Endosternite.

Endoskeleton

The endoskeleton of prosoma and mesosoma is present in the form of cartilaginous plate called *endosternite*. It is places above the nerve cord and alimentary canal. It possesses apertures for the passage of the instestine and aorta. Its edges are produced into *anterior, posterior and lateral processes* for the attachment of muscles which control the movements of appendages. Another similar skeletal piece lies ventral to the nerve cord in the segment which bears pectines. The body of endosternite is unites posteriosly with a chitinous plate, called *diaphragm* which possesses apertures for the passage of nerve cord, intestine dorsal aorta and muscles.

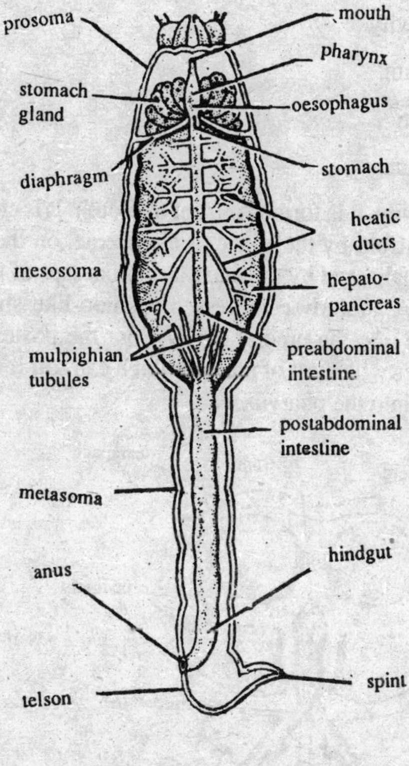

Fig. 17.7. Scorpion. AlimentaryCanal

Musculature. There is one pair of muscles lies in cephalothorax attached to it by one end and by other to the endosternite above nerve cord. In the pre-abdomen there are 8 pairs of dorso ventral muscles on the left and right of midgut, running from the dorsal to the ventral surface of the body.

Digestive System

The digesative system comprises the alimentary canal and digestive glands.

I. *Alimentary Canal.* The alimentary canal is complete and tube like nearly uniform. It is differentiated into four district regions.

1. Pre oral cavity
2. Stomodaeum
3. Mesenteron
4. Proctodaeum

1. *Preoral cavity.* It is formed infront of mouth. It is enclosed above by the rostrum, infront by the coxae of chelicerae, on the sides by the coxae of the pedipalpi and by the maxillary processes of first two pairs of walking legs. A laterally compressed cushion-like structure called rostrum protrudes into the preoral cavity from the posterior side. The rostrum is provides with a set of muscle which can roll it into a tube for sucking the fluid into the pharynx.

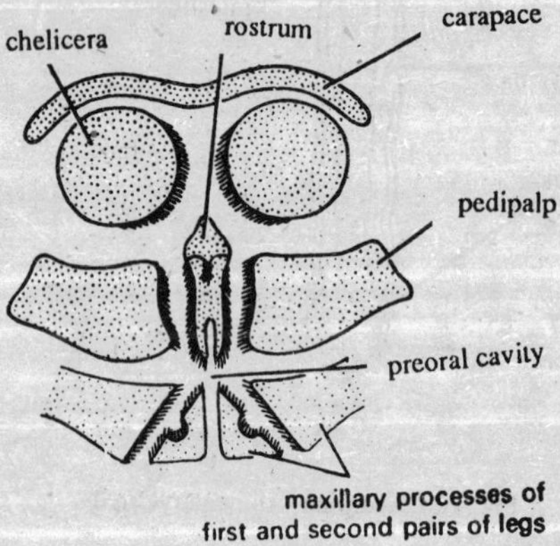

Fig. 17.8. Scorpion. T.S. through preoral cavity.

2. *Stomodaeum or foregut.* It includes mouth, phanynx and oesophagus and is lines internally by the cuticle.

(i) *Mouth.* It is a small narrow and transverse aperture just behind the preoral cavity. Only juces and pulps can enter the narrow mouth. Mouth leads into pharynx.

(ii) *Pharynx.* The *pha*rynx is a large, pear-shaped and muscular structure situated obliquely below the mouth. Due to the presence of several raadiating bundles of muscles radiating outwards from pharyngeal wall to the wall of the cephalothorax, its walls are highly elastic and capable of great dilation. The muscles serve the pharynx to act as a sucking organ so that the liquid food is sucked in through the mouth. The pharynx is followed by oesophagus.

(iii) *Oesophagus.* It is a short, narrow tube that passes through the nerve ring. The oesophagus extends into the midgut to from the *sleeve valve* which prevents regurgitation of food. It leads into stomach.

3. *Mesentron or midgut*. It includes stomach and intestine. These are lined with epithelium.

(i) *Stomach.* The stomach is a short thin walled and dilated tubul r structure situated in the cephalothorax. It extends upto the dia phragm. A trilobed, brownish gland, lying in the cephalothorax opens into the stomach. *Huxley* regarded it as *salivary gland* but *Blanchard* and *Pavlovsky* called it *stomach gland.*

(ii) *Intestine.* It is the longest part of the alimentary canal and extends upto the last abdominal segment where it joins the hindgut. The intestine is wide tube-like structure with glandular wall. It is divisible into two parts: *Preadominal* and *postabdominal*. At the junction of these two is a constriction and from here one or two pairs of Malpighian tubules arise. These tubules are excretory in function.

4. *Hindgut or proctodaeum.* It is the smallest part of the alimentary canal, restricted to the last metasomatic segment. it is internally lined by chitin and opens to the exterior through the *anal aperture* situated ventrally at the base of the stinging apparatus.

II. *Digestive glands*. There are a pair of *stomach glands* and a large *hepatopancreas*.

(i) *Stomach - gland*. It is a trilobed brownish gland and open into stomach. It secretes digestive enzymes like pepsin, lipase, eraspin etc.

(ii) *Hepatopancreas*. It is a large, brownish and lobulated gland occupies the entire preabdominal cavity. It has a mid-dorsal longitudinal groove in which lodges the heart. Intestine and other organs are found embedded in it. The wall of the lobules is composed of a layer of epithelium arranged on a thin membrane or *tunica properia* and a peritoneal covering. Two types of cells are found in the epithelium (a) large numerous *absorptive cells* and (b) fewer smaller *ferment cells*. Hepatopancreas sends 5 pairs of ducts open into the stomach.

Food and feeding. Scorpions are carnivorous and predaceous animals. They also practise cannibalism. They feed upon small animals like insects and spiders. The prey is sized and held by powerful chelate pedipalpi or being paralysed by the sting if the prey is large. The pedipalpi pass the killed prey to chelicerae which tear it to pieces. The food is then transferred to the pre-oral cavity which helps in squeezing the prey and oozing liquids are sucked in by the muscular pharynx. The process of feeding is very slow as the scorpion takes at least two hours to devour a cockroach. Scorpions can live without food for months together (about 6 months).

Digestion. Inside the preoral cavity the food is reduced to a pulp. Its protcins are partially digested by the secretion of certain *alveolar glands* which are present in the maxillary processes of first two pairs of legs. The food is reached to the stomach in partly digested state. Here the food is mixed with the secretion of stomach gland. The secretion contains pepsin, erepsin and lipase enzymes. Most of the digestion completed in stomach. In stomach hepatopancreas pour its secretion that contain anylase, lipase and proteinases enzymes. The digested food is absorbed into the absorptive cells of the hepatopancreas for storage. The food is sent to the intestine where water and remaining digested food material is absorbed. The undigested food is sent into the hindgut from where it goes out through the anus.

Circulatory System

In scorpions, the circulatory system is *open* type. It consits of:

1. Heart

11. Arteries

111. Sinuses

IV Pulmonary Veins

V Blood

1. *Heart.* The heart is an elongated and muscular tube of a faint greenish colour. It lies in the pre-abdominal region, just beneath the terga, in a deep median dorsal groove of the heptopancreas. The heart is divided from outside into seven chambers by shallow tranverse constrictions. The heart has seven pairs of dorso-lateral and valvular apertures or *ostia,* each chamber possessing one pair.

The heart lies suspended by means of ligaments in a thin-walled, membranous sheath, the *pericardium*. Ligaments divide the pericardial cavity into four compartments: one dorsal, one ventral and two lateral.

II. *Arteries.* The heart is continued anteriosly as *anterior aorta* and posteriorly as *posterior aorta.* A series of paired lateral *systemic arteries* are also given off from the heart.

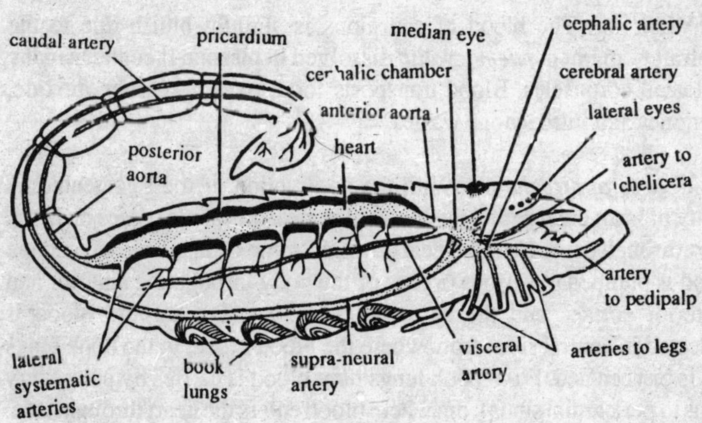

Fig. 17.9. Scorpion. Circulatory System

1. *Anterior aorta.* From anterior aorta a pair of *visceral arteries* are given off to the hepatopancreas and the intestine. After piercing the diaphragm it expands to form a small chamber behind the brain. The arteries to all the cephalic and thoracic appendages are given off from this chamber. It also gives off a pair of *ventro lateral* arteries which curves round the oesophagus and unite to form the *supra-renal artery.* This artery runs behind above the ventral nerve cord and supplies posterior region of the body.

2. *Posterior aorta.* It is also known as *caudal artery* and arises from the posterior end of the heart. It lies on the dorsal side of intestine and supplies blood to the intestine, muscles, telson etc. It ends in poison glands.

3. *Systemic arteries.* A systemic artery is given of on either side of each chamber of the heart. They ramify from a networks and supply the blood to different organs found in the abdomen.

III *Sinuses.* The ultimate branches of arteries open into blood spaces or lacunae among the viscera. The blood from the lacumae is collected into five larger *sinuses.* Out of these, a *pericardial sinus* surrounds the heart, *drosal sinus* above pericardial sinus, two *lateral sinuses* are present on lateral sides and a large *ventral sinus* is situated on the ventral side.

V. *Pulmonary veins.* The oxygentated blood from each book lung is carries by a pulmonary vein to the pericardial sinus.

VI *Blood.* The blood of scorpions is slightly bluish due to the respiratory pigment *haemocyanin* dissolved in plasma. It contains many nucleated corpuscles. Blood transports food, oxygen, carbon dioxide, hormones and nitrogenous wastes.

Course of circulation. With the contraction of the ligaments, the cavity of heart enlarge and the blood of pericardial sinus passes on to the heart through ostia. Now the heart contracts, the ostia are closed and the blood is pumped to various organs of the body through the anterior and posterior aortae and the systematic arteries. The impure blood is collected in ventral sinus from where the blood is sent to the book lungs and is oxygenated. From book lungs pure blood is carries by pulmonary veins to pericardial sinus from where blood enters the heart through ostia.

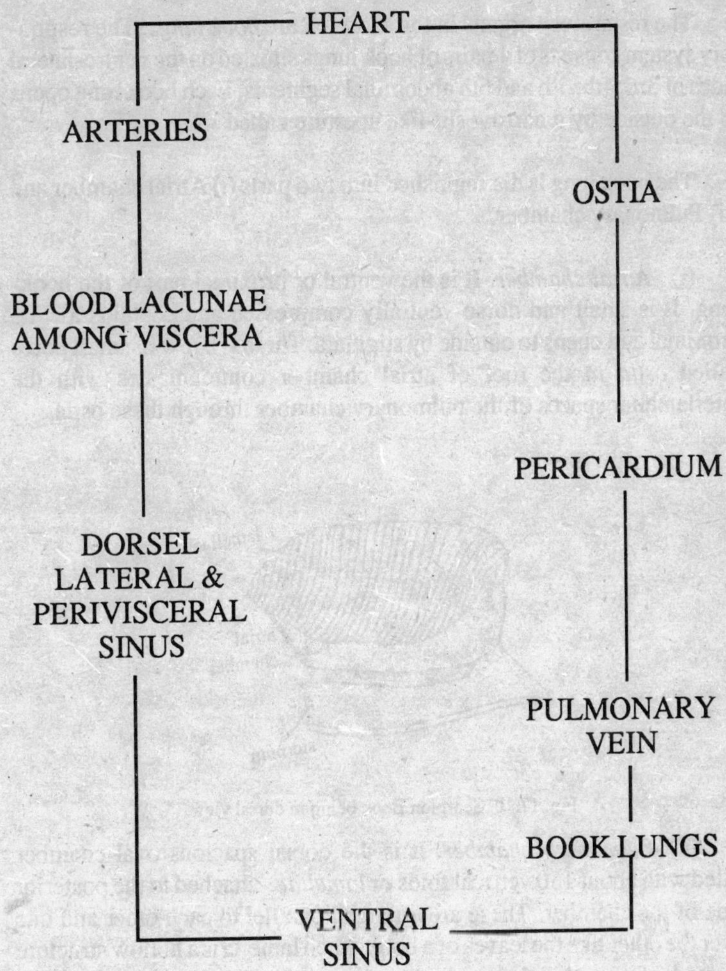

Chart Showing Course of Circulation in Scorpion

Respiratory System

The respiratory organs in the scorpion are *book lungs*. The respiratory system consists of 4 pairs of book lungs situated on the ventro-lateral sides of 3rd, 4th, 5th and 6th abdominal segments. Each book lung opens to the outside by a narrow slit-like aperture called *stigmata*.

The book lung is distinguished into two parts (i) Atrial chamber and (ii) Pulmonary chamber.

(i) *Atrial chamber*. It is the ventral or proximal part of the book-lung. It is small and dorso-ventrally compressed and contains air. Its proximal and opens to outside by stigmata. There is a row of small pores called *ostia* in the roof of atrial chamber communicates with the interlamellar spaces of the pulmonary chamber through these ostia.

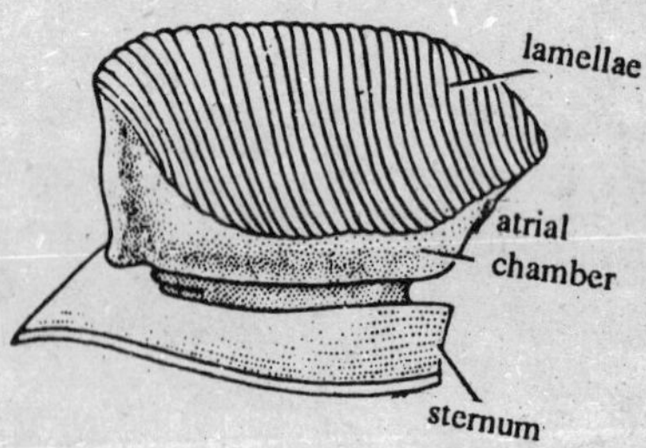

Fig. 17.10. Scorpion Book being in dorsal view.

(ii) *Pulmonary chamber*. It is the dorsal spacious oval chamber filled with about 150 vertical folds or *lamellae*, attached to the posterior side of the chamber. These are arranged parallel to each other and one over the other like the leaves of a book. Each lamella is a hollow structure formed of two thin layers of cuticle unites at their edges and thus enclosing *interlamellar spaces* filled with blood. The adjacent lamellae enclose *interlamellar air spaces* filled with air. The interlamellar air spaces communicate with the atrial chamber through ostia.

Blood supply. The deoxygenated blood is sent to each book-lung from the ventral sinus by *diverticulum*, from where the blood gets filled

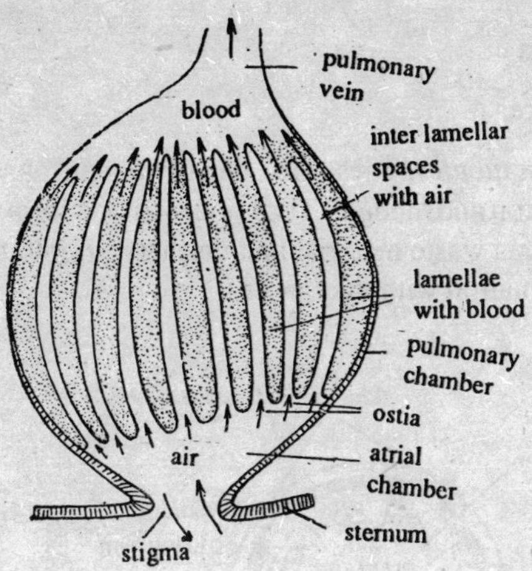

Fig. 17.11. Scropion. V.S. of book ıung.

in the intralamellar spaces of the lamellae. The aerate blood from each book-lung is drained by a *pulmonary* vein to the pericardium.

Mechanism of respiration. Air is drawn into and expelled out from the book-lungs by the action of dorso-ventral and atrial muscles. Contraction of these muscles compresses the book-lungs this expells out the foul air of the interlamellar spaces into the atrial chambers and than to the ourside through the stigmata- when these muscles relax, the pulmonary sac regains their normal shape as a result of which the fresh air rushes the atrial chambers throught the stigmata and then fills the interlamellar spaces.

The lamellae contain deoxygenated blood and are surrounded by fresh air. Thin walls of the lamellae permit the exchange of gases by diffusion. The blood takes up oxygen from and gives up carbon dioxide to the air. The oxygenated blood and befouled air are then replaccs and process is repeated.

Excretory System

The excretory system of scorpion consists of

(i) Malpighian tubules

(ii) Coxal glands

(iii) Hepatopancreas

(i) *Malpighian tubules.* These are one or two pairs attached to the junction of mesenteron and hindgut. They float in blood from which they eleminate nitrogenous waste material and discharge it into the alimentary canal for elemination with the faeces.

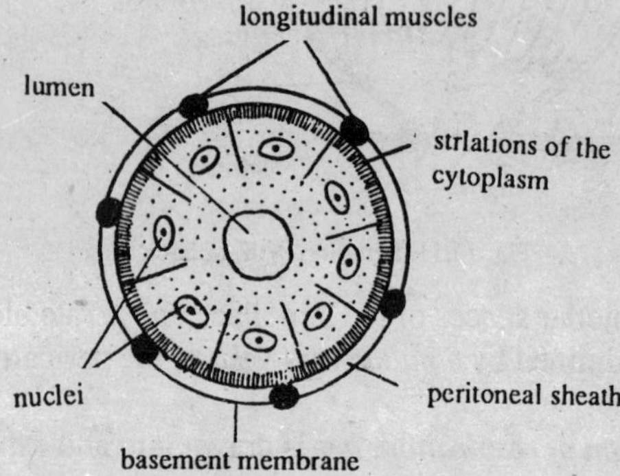

Fig. 17.12. Scorpion. T.S. of Malpighian tubule.

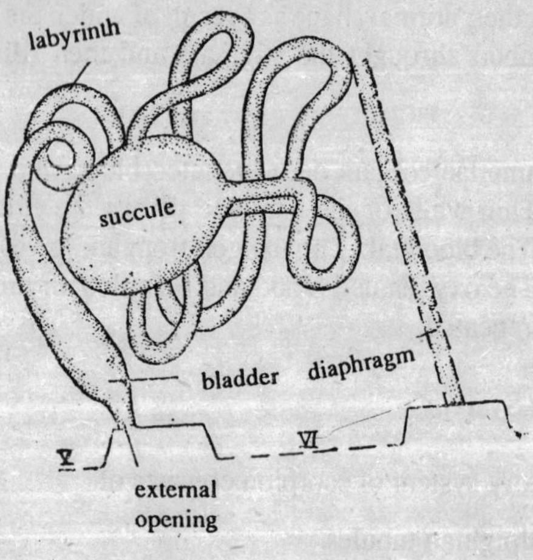

Fig. 17.13. Scorpion. Coxal gland.

(ii) *Coxal glands.* A pair of coxal glands is situated near the base of the 5th walking leg. Each consists of three parts: a large *saccule* or *end sac,* a long coiled duct or *labyrinth* and a small terminal *bladder* or reservior. The bladder opens outside by means of a minute pore situated on the posterior surface of the coxa of the fifth walking leg.

The coxal glands of scorpion are considered homologous with the antennary glands of prawn. The saccule and tubule of each coxal gland collect excretory nitrogenous wastes from blood and pass it out side through the excretory pore. Urate crystals are notices in the saccule.

(iii) *Hepatopancreas.* According to *Pavlovsky* the hepatopancreas of scorpions also serves as an excretory organ. Accroding to him if ammonium carmine is injected in the body cavity of scorpion is collected in the form of small, bright, red granules by the cells of hepatopancreas.

(iv) *Nephrocytes.* Large nephrocytes and lymph tissue organs, present beneath the body wall in mesosoma, are both excretory as well as phagocytic in nature.

Nervous system

The nervous system consists of:

1. Brain or Supra-oesophageal ganglion
2. Circum-oesophageal connectives
3. Sub -oesophageal ganglion
4. Ventral nerve cord

The brain is small bilobed mass situated in the prosoma just beneath the median eyes. From the brain arises a pair of thick, short and stout circum- oesophageal connectives which encircle the oesophagus and unite ventrally in a suboesophageal ganglion. The suboesophageal ganglion is continued backwards into the abdomen as double ventral nerve cord. It extends upto the 4th segment of the postabdomen. The nerve cord bears three segmental ganglia in the preabdomen and four in the postabdominal segment. The nerve cord is rounded and slender in the preabdomen but flattened and ribbon shaped in the postabdomen.

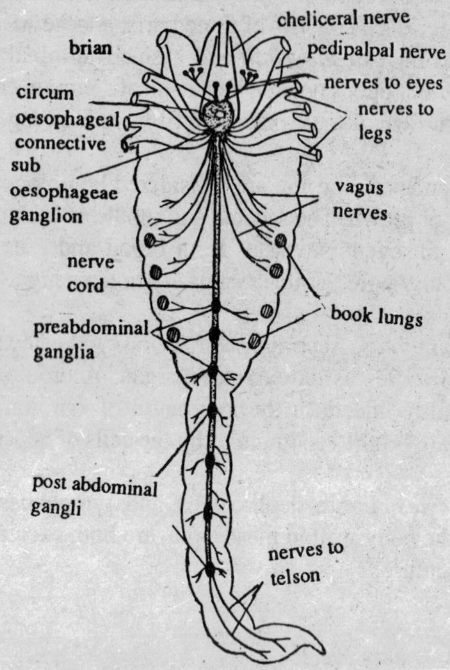

Fig. 17.14. Scorpion. Nervous system

Brain gives off a pair of *optic nerves* to the marginal eyes. Besides, the brain gives off numerous median delicate nerves to the rostrum, pharnyx and oesophagus. From the cerebral ganglion and circum oesophageal connectives arise six pairs of *lateral nerves*. These nerves innervate the six pairs of cephalothoracic appendages. The sub-oesophageal ganglion sends two to four pairs of *vagus nerves*. These run posteriorly into the preabdomen and supply the genital operculum, pectines and the first two pairs of book-lungs.

The segmental ganglia of the preabdomen sent out nerves to the last two pairs of book-lungs, heart and muscles of the abdomen. The segmental ganglia of the postabdomen supply nerves to the segmental muscles. The last postabdominal ganglion also supplies the sting.

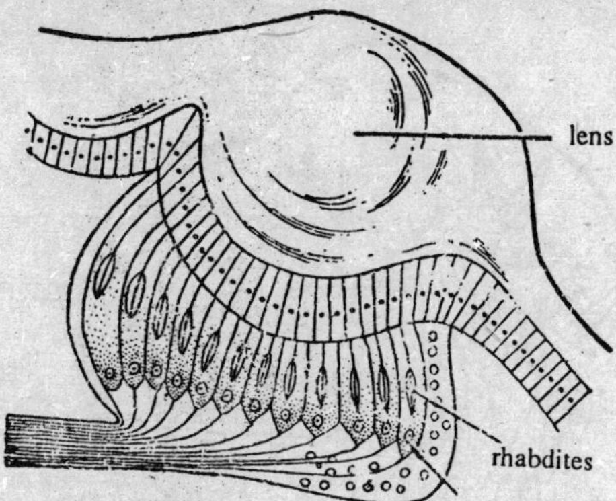

Fig. 17.15. Scorpion. V.S. of median eye

Sense Organs

The sense organs of scorpion include the eyes and pectines .

1. *Eyes*. There are four pairs of eyes in scorpion - one pair of median eyes and three pairs of lateral eyes.

(i) *Median eyes*. Median eye is like a cup covered externally by a cuticular *lens* which is continuous with the cuticle but is much thicker. Inside the pigmented cup are *rhabdomes,* each enclosed inside several retinal cells which receive nerve fibres of an optic nerve.

(ii) *Lateral eyes*. These are similar to simple eyes of insects. These are like a pigmented cup covered externally by a biconvex lens formed from transparent cuticle. Within the cup are several longitudinal optic rods called *rhabdomes* associated with *retinal cell* or *retinulae*. The retinel cells receive nerve fibres.

2. *Pectines*. These are a pair of comb-like structures borne on the sternum of 2nd preabdominal segment. Their process have many sensory cells. They are tectile organs. They are also considered to be sensitive to the ground vibrations and changes in humidity.

It is also regarded that scorpions have some preception of sound. In some species of scorpions *stridulating organs* are found on the coxae of

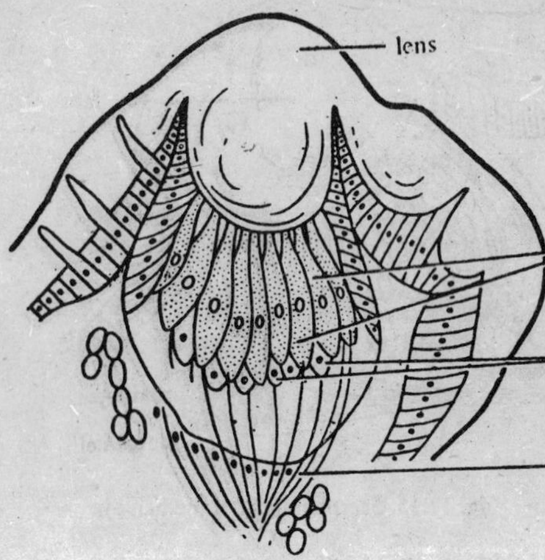

Fig. 17.16. Scorpion. V.S. of lateral eye.

pedipalpi in the form of ridges, across which file-like surface can be drawn to produce some sort of sound.

Reproductive System

The sexes are separate. Externally male and female are alike but they can be distinguished on the basis of following characters:

(i) The female is generally larger than the male.

(ii) The abdomen of female is broader than that of the males.

(iii) The pectines of female are smaller than of male.

(iv) The pedipalpi of female are smaller than those of male.

1. *Male reproductive system.* The male reproductive system comprises.

1. Testes
2. Vasa deferentia
3. Seminal vesicles
4. Accessory glands

5. Genital chamber

6. Paraxial organs.

1. *Testes*. A pair of tests are long mesh-like structure which lie embeded in the hepatopancreas from 3rd to 6th segment of mesosoma. A testis consists of two slender longitudinal tubules connected together by four transverse tubules so as to form three squares. Each tubule is lines with a layer of germinal epithelium. The cells of epithelium divide to form sperms which are filiform and motile.

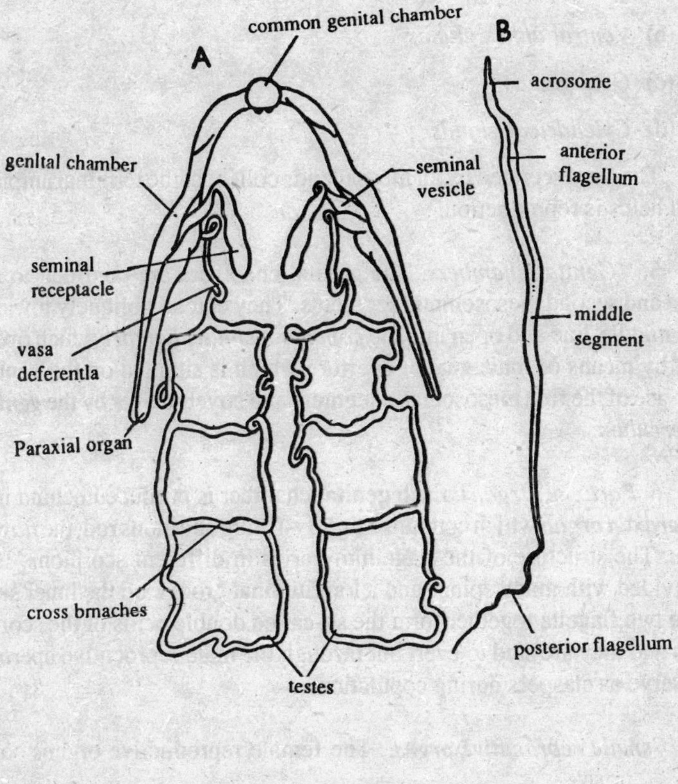

Fig. 17.17. Scorpion. A = Male reproductive organs, B = A sperm.

2. *Vasa deferentia.* Anteriorly, a small thin duct arises from the other end of each testis. It runs forward and outward to atrium. Prior to its opening, it widens and this dilated part is known as *terminal ampulla* of vas deferens. The terminal ampulla receives the opening of accessory glands and the seminal vesicles.

3. *Seminal vesicle* - It is a small club-shaped structure which extends backwards from the terminal ampulla of vas deferens and opens into it. It acts as an organ of storage of mature spermatozoa.

4. *Accessory glands.* Four pairs of accessory glands are associated with the male reproductive organs These are:

(a) *Dorsal annex glands*

(b) *Ventral annex glands*

(c) *Oval glands*

(d) *Cylendrical glands*

The fluid secreted by all these glands collect in the terminal ampulla and helps is reproduction.

5. *Genital chambers.* The genital chambers are confined to the first and second mesosomatic segments. They extend obliquely towards the middle-line and open into the *common genital chamber* which opens out by means of male genital aperture which is situated on the ventral surface of the first mesosomatic segment and covered over by the *genital operculum.*

6. *Paraxial Organs.* Each genital chamber is produced behind into a *paraxial organ* which contains a tightly-fitting chitinous rod, the *flagellum.* The structure of the flagellum varies in different scorpions. It is provided with many spines and a longitudinal groove on the inner side. The two flagella together form the so-called double penis of the scorpions and they are said to evert out through the male reprocutive aperture to serve as claspers during copulation.

II. *Female reprocutive orgns.* The female reproductive organs comprise:

1. Ovary
2. Oviducts
3. Common genital chamber

1. *Ovary.* There is a single ovary in scorpion which extends from the 3rd segment to the 6th segment of the mesosoma. It remains embeded in the hepatopaucreas. It consists of three long and narrow tubules called *ovarioles.* Of them the middle one is smaller and called *median ovariole.* The other two are long and present on the lateral sides and called *lateral ovarioles.* All the ovarioles are connected with each other at four places by *transverse connectives* giving a ladder like appearence. All the ovarioles and tranverse connectives are lines with germinal epithelium, the cells of which divide to form *ora.* Many hollow sac-like structures, called *follicles,* arise from the ovarioles and the transverse connectives. Each follicle is male up of two parts a proximal swollen part, the *base* and a narrow distal part, the *appendix.* The tip of appendix enlarges to form the *head.* The development of embryos takes place within these follicles hence they ate considered equivalent to the uteri of other animals.

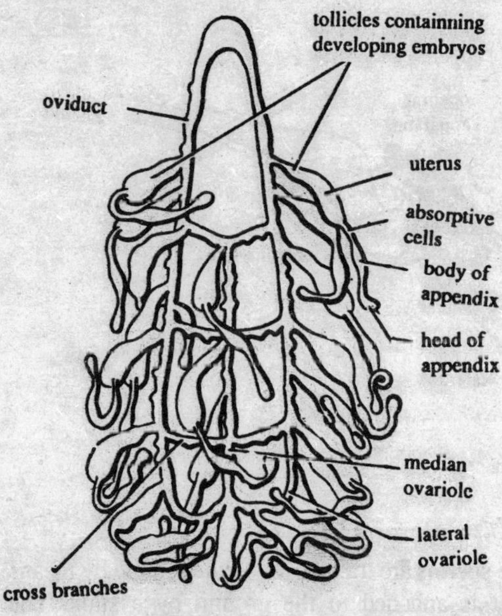

Fig. 17.18. Scorpion. Female reproductive organs.

2. *Oviducts.* A pair of tubes, the *oviduct,* starts from the anterior ends of the lateral ovarioles. These extend forwards for a short distance and then converge to open into a median sac, the *genital chamber*

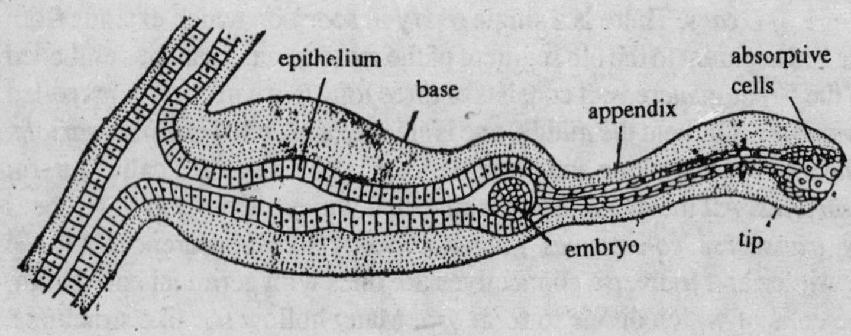

Fig. 17.19. Scorpion. diverticulum of ovarian tubule

3. *Genital chamber.* The genital chamber is situated in the first mesomatic segment. It opens out side by the *female genital aperture* situated on the ventral side of first mesosomatic segment being covered by the bifid *genital operculum.*

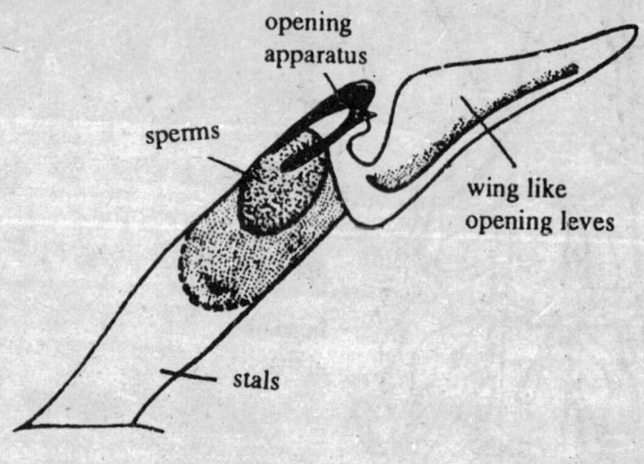

Fig. 17.20. Scorpion: A spermatophore

Spermatophore. The sperms are transferred to the female by means of a *spermatophore* that is attached to the ground by a stalk. The spermatophore bears a wing like lever at its free end. Connected to the lever is a pocket carrying sperms and also an opening apparatus. As the lever is pressed the opening apparatus is released and the sperms are triggered into the female genital apertuse.

Copulation and fertilization. Fabre has described the mating habits of scorpion. Male and female face each other, raise their postabdomens high into the air, and move about in cricles. The male then holds the pedipali of the female with his own peidpalpi and both interwine their postabdomens above them. They now walk together forward and backward and in rings. While moving in a ring, the male retreats, dragging the female after it. Courtship may last for hours or even days. Finally, the male digs a burrow and both enter it and mate. The male is often eaten up by the female after mating.Eventually the male deposits a spermatophore, that is attached to the ground. The male manoeuverrs the female so that the genital area is over the spermatophore. The sperms enter into the female orifice and fertilization takes place. The Indian scorpions breed during monsoons.

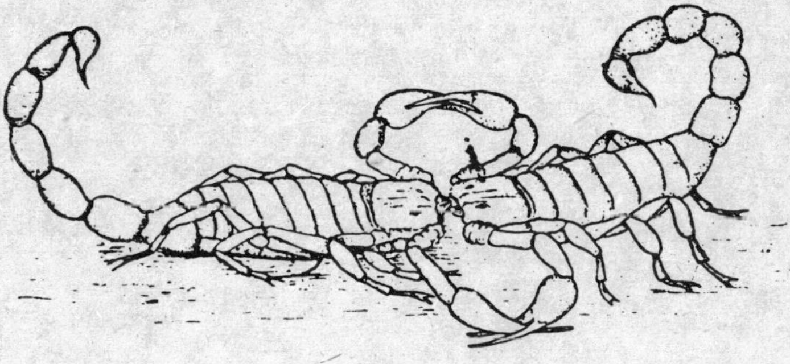

Fig. 17.21. Copulation in Scorpious.

Development. Development takes several months or cven a year or more. The scorrpions are *viviparous* and the mother gives birth to two or three dozens of young ones at a time. The egg is oval or spherical which is rich in yolk and is surrounded by a thin membrane lies in a follicle that ariscs as outgrowth of the walls of the ovariole or transverse connective.

After birth the young ones are carried by the mother scorpion on her back for some days till they can care themselves. Young ones moult several tines before they grow into adults. Young ones than leave the mother and then become independent.

18

PERIPATUS

Peripatus or the "walking worm" belong to the small phylum *Onychophora*. Its position has been controversial in some texts Onychophora has been ranked as class where as in others as subphylum. In modern texts it is treated as a separate phylum that includes only about 70 living species. It has characteristics of both the Annelida and Arthropoda and is referred to as the "connecting link".

Historical account

Phylum Onychophora is represented by a single genus, *Peripatus* which was wrongly placed in phylum Mollusca by *Guilding* (1826) because of its slug-like appearence. But soon after the discovery of tracheae in the animal by *Mosely* (1874) it was classified as a class of phylum Arthropoda. *Sedgwick* has given a monograph an *Peripatus*.

Geographical Distribution

Peripatus is practically having its wide range of distribution both in tropical and subtropical areas. As it is shown in the table below, it is one of the best example of *discontinuous distribution*.

No.	Geographical distribution	Nomenclature given by Sedgwick	Name	No.of species
1.	Maxico and W.Indies.	*Neoperipatus*	*Peripatus*	29
2.	W. Africa and Congo.	*Congoperipatus*	*Mesoperipatus*	1
3.	Malaya		*Eoperipatus*	4
4.	S. Africa, Netal to Cape	*Capoperipatus*	*Periperipatus*	7
5.	Malaya Penninsula	*Malayan Peripatus*	*Peraperipatus*	1
6. 8.	Australia, Tusmania, New Zealand.	*Austroperipatus*	*Parapaetoides*	
7.	Chili Island	*Chiloperipatus*	*Chiliperipatus*	1
8.	Foot hill of Himalyas		*Yphelloperipatus*	1

Habits and Habitat

Peripatus is exclusively a terrestrial and walking form and living in a demp leaf moulds, barks of trees, under the stones, logs and in the cervices of rocks, but not in salt water medium.

They are nocturnal in habit, usually they come out at night for feeding. They are carnivorous and capture prey by shooting out a glue-like substance from openings on either side of the mouth. Small animals such as crickets and spiders quickly become entangled in the sticky glue.

External Features

These individuals possess a reddish colour on the ventral side, while darker hues on the dorsal side. The body is cylindrically elongated, bilaterally symmetrical and small. It measures 1.5 - 15 cm in length, according to species. The skin is velvety, and thrown into transverse ridges bearing number of wart-like papillae furnished or armed into

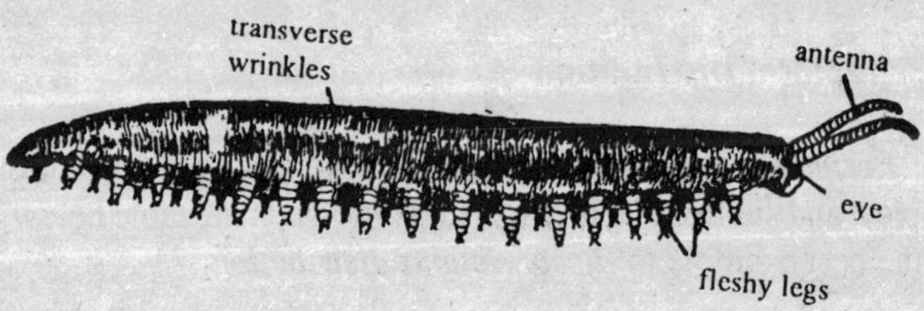

Fig. 18.1 *Peripatus*.

chitinous spines. Externally the body is not much marked segmented but each possess a pair of appendages.

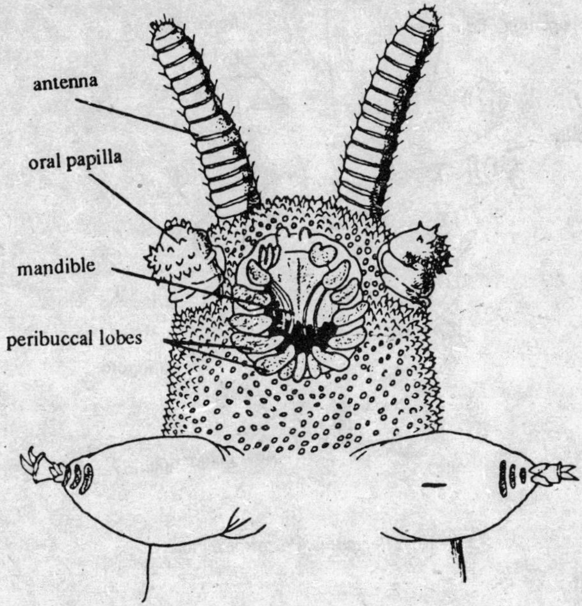

Fig. 18.2. *Peripatus*. Anterior region

First three segments form *head*, which is not distinct. It bears a pair of segmented antennae, a pair of eyes situated dorsally. Mouth is mid-ventrally situated bearing a pair of small horny jaws surrounded by a circular lip or *circumoral fold* and a pair of blunt *oral papillae* on either sides. Each oral papilla bears an aperture for slime gland which produce a sticky glue-like secretion for catching prey.

A series of short, stumpy, conicasl legs arise ventro-laterally from the trunk. All the appendages are alike but for the last 2 or 3 are reduced. Their number varies from 14-43 pairs from one species to another. Each leg is short, unsegmented having tubercle and spiny pad on the ventral side. Each leg is hollow with terminal foot bearing a pair of curved claw.

Genital opening always lies infront of the *anus* which is situated at the posterior end of the body.

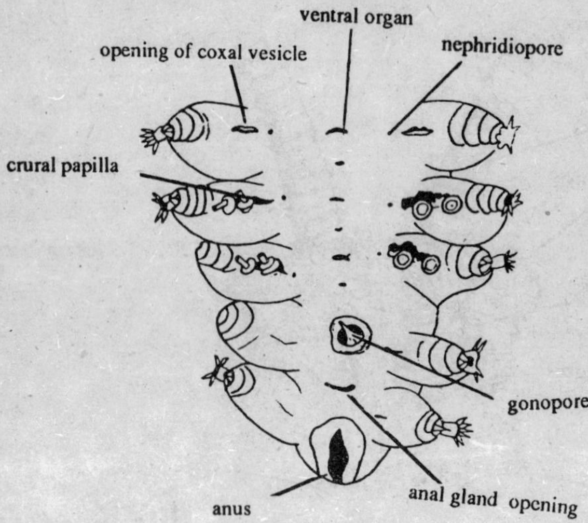

Fig. 18.3. *Peripatus*. Posterior region

Body Wall

The body wall is of annelidan type so the skin is formed of three layers, *cuticle, epidermis* and *dermis*. The exoskeleton is absent. The *cuticle* is thin flexible, chitinous and waterproof covering. It bears numerous velvety processes. It is secreted by cuticle. Beneath the cuticle is a single layer of epidrmis, a thin basement membrane. The dermis is made up of connective tissue and there is a complex musculature composed of circular, longitudinal and transverse muscle layers. Beneath dermis is epithelial layer which lines the coelom or body cavity and invests the contained organs. By means of incomplete muscular partition the body cavity is divieded into a median and two lateral compartments.

Digestive System

The alimentary canal is simple and straight as in an annelid but the buccal cavity is formed secondarily by the union of a papillary ring and folds surrounding the true *mouth* into a circular lip. It contains a tongue and the jaws. Into the posterior end of buccal cavity there is a median

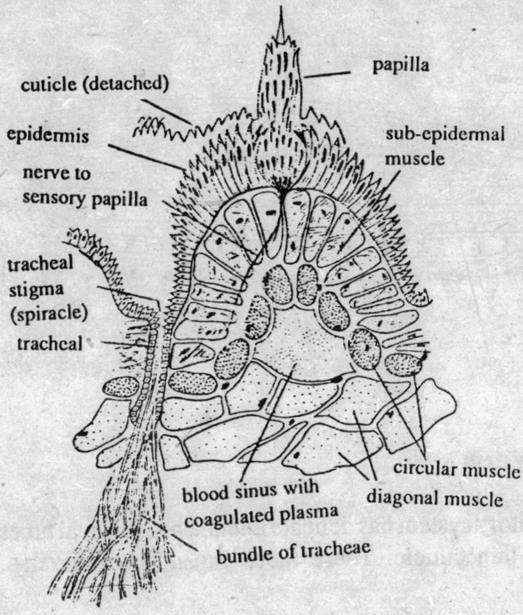

Fig. 18.4. *Peripatus*. Section of skin

opening of salivary glands. The saliwary glands are elongated, narow and tubular, representing modified nephridia. Buccal cavity follows a thick, muscular *pharynx*. Pharynx is follwed by *oesophagus* a wide *midgut* and a short hind gut or *rectum*. The rectum opens out through the *anus*.

A Pair of elongated and branches *slima glands* are present in the coelom. These open on the tips of oral papillae and secrete a glue-like adhesive substance which hardens on contact with the air. This secretion helps in catching the prey. Digestion is inter-cellular takes place in midgut.

Circulatory System

The *heart* is an elongated muscular, mid-dorsal, contractile tube lying in the *pericardial sinus*. It is open at both the ends and bears several pairs of lateral ostia, Blood flows forward into various organs, there are no blood vessels to return the blood, the blood in collected into various sinuses and than pass to the heart.

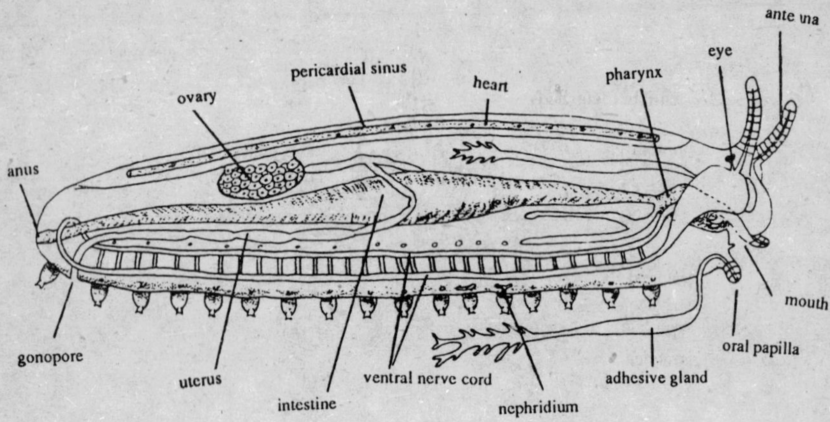

Respiratory System

The respiratory system has delicate unbranched tracheal tubes internally lines by thin cuticle. These tubes expand in transverse plane

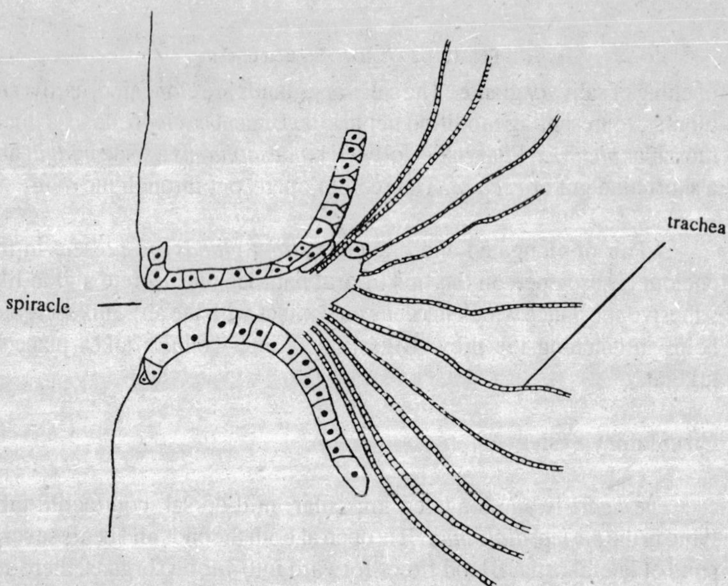

Fig. 18.6. *Peripatus*. Spiracle and Trachea.

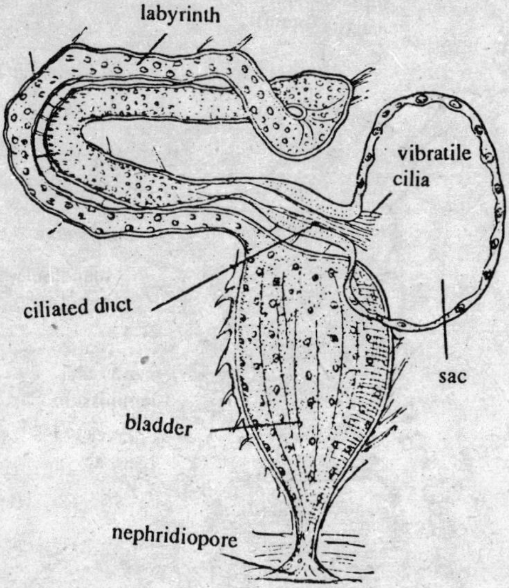

Fig. 18.7. *Peripatus. A nephridium*

through out the body and open externally by small pores called *stigmata* or *spiracels* . The stigmata lie in little dipressions of the integument, the *tracheal pits.* The number of spiracels varies from species to species, through these the air is drwan into the body.

Excretory System

The excretory system is annelidean type which is effected by the segmentally arranged *nephridia.* A nephridium consists of distally a *bladder* extinding into a *coiled tube*. It opens externally by *nephridi opore* situated at the base of leg. Internally it opens into colemic sac by a *ciliated funnel*. The bladder is originated from the ectoderm and the rest of the part from mesoderm. Other than nephridia there are *coxal* or *crual glands* lying in the lateral component of the body cavity whose ducts open on the lower surface of the legs just outside the nephridiopore. These glands are absent in females. The *Peripatus* is *uricotelic,* excreting chiefly uric acid as nitrogenous waste product.

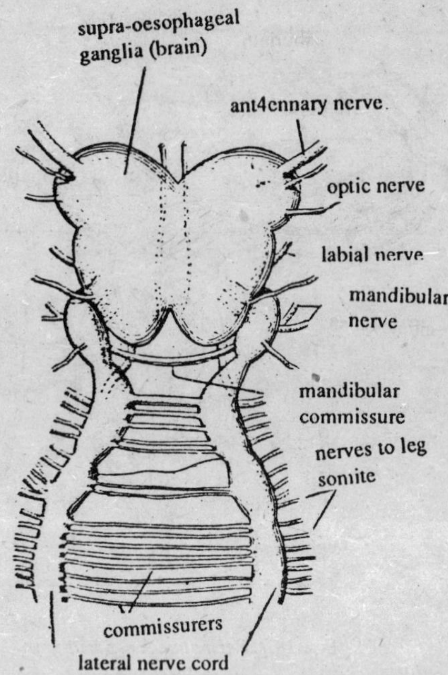

Fig. 18.8. *Peripatus. Brain*

Nervous System

The nervous system is primitive and is similar to that of annelids. It consists of two oval ganglia, the *brain* in the anterior part of the body lying over the pharynx. Two parallel longitudinal ventral nerve cords joined by many commissures assuming ladder-like in appearence.

Eyes.

The eyes are typical, resembling those of annelids in all respects. It is obvious when the animal is irritated their secretion is discharges in the form of fine viscid thread which might be the mesure of defence.

Reproductive System.

The males are smaller than females. Sexes are separate and the reproductive ducts are ciliated. The gonads are paired and lie posteriosly aabove the intestine.

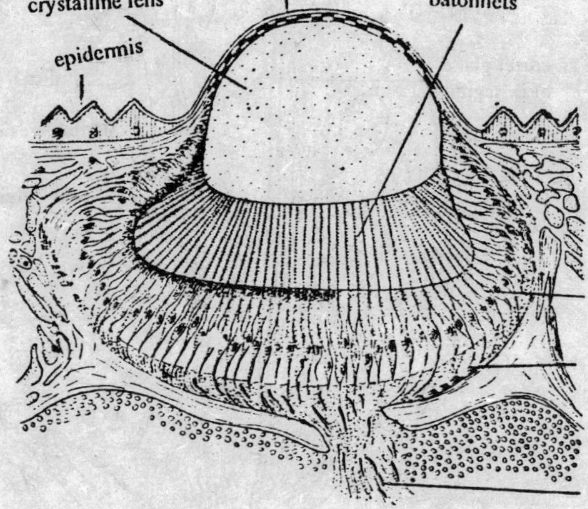

Fig. 18.9. ***Peripatus.*** **V.S.of eye.**

(i) *Male reproductive organs.* A pair of *testes* which each externd into a narrow tube -like *vas eferens.* Both the vasa efferentia open into bulb-like structure, the *seminal Vesicles.* Each vesicle is followed by *vas deferens.* The vasa deferentia are coiled structures and both unite to form an *ejaculatory duct.* The ejaculatory duct opens out through male genital pore. In connection with the reproductive system are *crural glands* found on all the legs except the first and are better developed in males. The sperms are glued together in ejaculatory duct to form *spermatophore.* The gonopore lies between the last pair of legs on ventral side.

(ii) *Female reproductive organs.* It comprises a pair of tubular *ovaries,* a pair of *oviducts* and two long *uteri.* Uteri unite behind into a median *vagina* opening into genital pore. Some species may have a *seminal receptacle* connected with the uterus and communicating with a funnel shaped *recepticulum ovorum* embracing the ovaries. In oviparous forms a long ovipositor is formed from the vagina.

Copulation. The real process of copulation is not yet observed and nothing is known about sperm enterance so far. The fertilization is

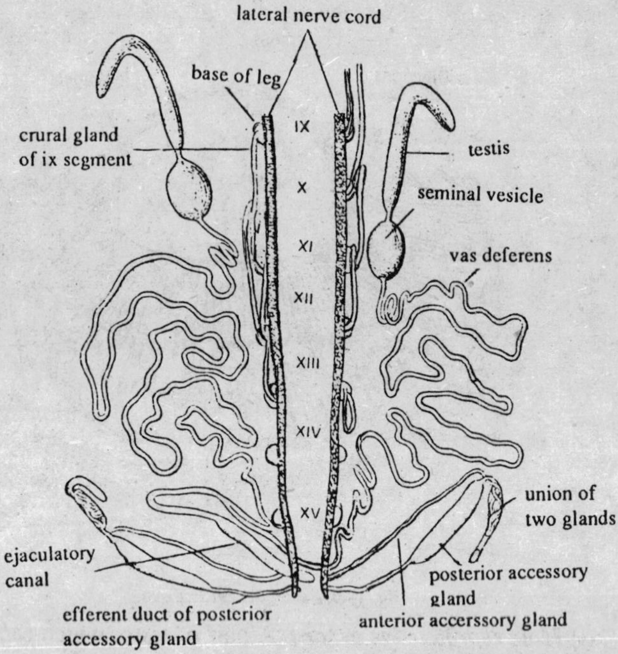

Fig. 18.10. *Peripatus*. Male reproductive organs.

internal sperms are not found in the uterus and oviducts so it appears probably that they reach directly to the ovary by boring through the skin or traversing the body cavity. Mostly they are *viviparous* but a few are *oviparous*.In viviparous forms embryos receive nourishment from the uterine wall through a *placenta* or they develop at the expanse of yolk. Oviparous forms lay eggs with advanced embryos. The young ones are alike to their parents except in size and colour. So that the mother does not appear to pay any special attention to her young ones, which wander away and get their own food.

Affinities

The *Peripatus* is very important from the zoological point of view because it displays some characteristics of annelids and some characteristics of arthropods along with some specialized features of its own. Because it shares the characters of two phyla, it is considered to be a

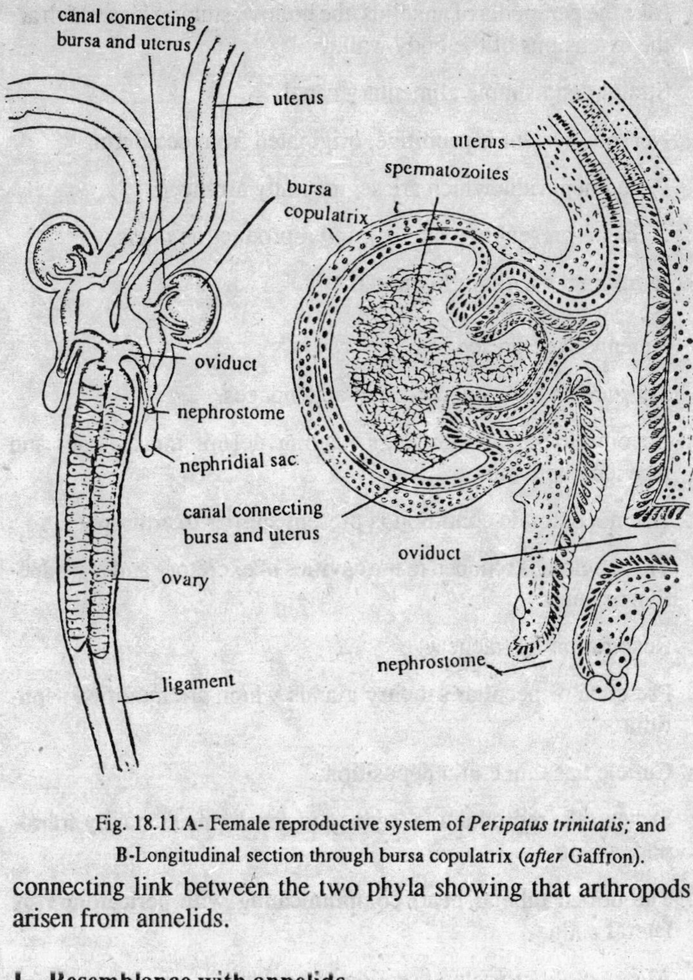

Fig. 18.11 A- Female reproductive system of *Peripatus trinitatis;* and B-Longitudinal section through bursa copulatrix (*after* Gaffron).

connecting link between the two phyla showing that arthropods have arisen from annelids.

I. Resemblance with annelids

1. Body is vemiform, with truncated extremities.
2. True head is absent.
3. Body wall consists of a thin cuticle beneath which presence of circular and longitudinal muscles.
4. Eyes alike and simple.

5. Like the parapodia of annelids, the hollow, stumpy legs which are the extensions of the body wall.
6. Straight and simple alimentary canal.
7. Nervous system is primitive, originated from ectoderm.
8. Paired nephridia which are segmentally arranged.
9. Cilia are present in excretory and reproduction ducts.

II. Resemblances with arthropods

1. Antennae are present.
2. Jaws are nothing but modified appendages.
3. Locomotion by definit legs, having definit musculature and provided with claws.
4. Haemocoelomic condition is present similar to arthropods.
5. True coelom is reduced to the cavities of excretory and reproductive organs.
6. Respiration by tracheae.
7. Presence of peculiar salivary glands which are modified nephridia.
8. Cuticle has thin chitin deposition.
9. Structure of reproductive organs and development manily arthropodan type.
10. The dorsal tubular heart communicating with pericardium by lateral ostia.
11. Absence of perivisceral section of coelom.

III. Its own Characters.

Peripatus exhibit certain characteristics of its own to recognize it as neither an annelid nor an arthropod.

1. Absence of external segmentation of the body.
2. Velvety and soft nature of skin.

3. To some extent the antennae of *Peripatus* can not be homologous to that of other arthropods.
4. Nature of 3- segmented head seems to be mid-way of arthropods (head in made up of 6 segments).
5. Restriction of jaws to a single pair.
6. Quite unevenly distributed spiracles and tracheae.
7. Absence of true ganglia, but separate two ventral nerve cords.
8. Difference in the distribution of reproductive organs.
9. Structure of eyes is less complicated.

Taxonomic Position

With the peculiar nature and some annelidan characters it appears that probably annelids is one of the source and that it is originated from the marine polycheat, which given up the marine habit and has become modified for locomotion on the land without jointed appendages. It seems to be entirely against to the rest of the arthropods, yet it has been kept under phylum Arthropoda, giving it an separate class rank.

According to *Lankester* (1904), of aquatic region, the Hyparthropoda (a hypothetical undefined group presumably without pre oral segments) were the first to appear from these arose the protarthropoda with one pre-oral segment, a terrestrial discendent having survived as *Peripatus*.

According to *Menton* (1950-52) linkage of this group Onychophora with two phyla will not be probably evidence of simiplification and moreover, the habitats of Onychophora provides ample reasons for the persistance of flexible body wall, leg like jaws together with the eloboration of a subcutaneous connective tissue instead of cuticular scutes.

According to *Snodgrass (1938) the* Onychophora and marine *Aysheaia* are possible discendantly from an extinct group of marine lopopod annelids and they may have no intimate relationship with trilobite chelicerate of precambrain arthropods. Onychophore bears chitinous mandibles which only might be preserved as fossil nothing of other parts known so far due to its soft nature.

No doubt the *Peripatus* is a living example and strictly terrestrail, and the only fossil refered to the class have been quesationed. Of these *Aysheaia* from middle cambrain is the best knwon fossil. It was a marine form, probabaly devoid of tracheae. Moreover, 10 pairs appendages but these bear with 6 claws, while the recent onychophores bear only two claws on their legs. It seems obvious that this *Aysheaia* may represent a very primitive onyhophore which is more closer to the annelids than the arthropods.

Apart from that it must have been derived archiannelidian stock because of the ventral nerve cords were primitive, widely separated conditions or it may more be correct to say that the present day annelids and *Peripatus* have got their origin from a common ancestor from which they independently evolved.

It is due to because that common features appeared between the onychophores and present day annelids are similar, being evolved from common ancestral stock in which such characters were probably present.

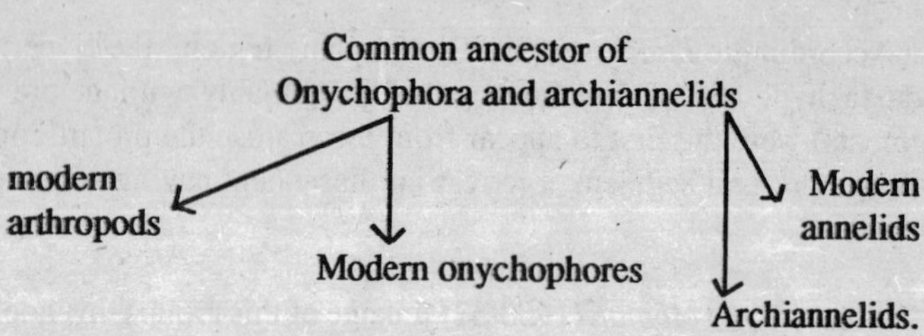

However, according to modern concept (based on *Barnes*, *Marshall* and *William*) this group was more wide spread and diversified. The view is supported by its discontinuous distribution and discovery of several fossils from middle cambrian. It is, therefore, believed that Onychophora is an ancient independent group of segmented animals contemporary to arthropods and as such has been placed under separate phylum onychophora.

INDEX